Permeable Pavements

Permeable Pavements Task Committee

EDITED BY

Bethany Eisenberg, LEED AP
Kelly Collins Lindow, PE
David R. Smith

SPONSORED BY

The Low Impact Development Committee of
The Urban Water Resources Research Council of
The Environmental and Water Resources Institute of
The American Society of Civil Engineers

PUBLISHED BY

The American Society of Civil Engineers

Cataloging-in-Publication Data on file with the Library of Congress

Published by American Society of Civil Engineers
1801 Alexander Bell Drive
Reston, Virginia, 20191-4382
www.asce.org/bookstore | ascelibrary.org

Cover photos (from top to bottom): Susan Lee of VHB; John T. Kevern, Ph.D., PE, LEED AP of University of Missouri-Kansas City; Hastings Pavement Company; and Kevin Earley of Anchor Block Company.

Contents

Acknowledgments

The ASCE/EWRI committee would like to acknowledge Vanasse Hangen Brustlin, Inc., (VHB) for allowing the use of figures and checklists that it developed. Contributing authors and VHB make no representation or warranty of any kind, whether expressed or implied, concerning the accuracy, completeness, suitability, or utility of any information including figures and checklists, apparatus, product, or process discussed in this publication, and assumes no liability therefor. This information should not be used without first securing competent advice with respect to its suitability for any general or specific application. Anyone utilizing this information assumes all liability arising from such use, including, but not limited to infringement of any patent or patents.

Task Committee Co-chairs

Bethany Eisenberg, LEED AP
VHB

Kelly Collins Lindow, PE
Cityscape Engineering, LLC

Technical Editors

Bethany Eisenberg, LEED AP
VHB

Kelly Collins Lindow, PE
Cityscape Engineering, LLC

David R. Smith
Interlocking Concrete Pavement Institute

Principal Authors

Eban Z. Bean, Ph.D.
Eastern Carolina University
(Chapters 8, 10)

Andrea M. Braga, PE, CPESC
Geosyntec Consultants, Inc.
(Chapters 2, 5)

Michael L. Clar, PE, DWRE
New Castle County, DE
(Chapters 1, 9)

Kelly Collins Lindow, PE
Cityscape Engineering, LLC
(Chapters 1, 6, 9, Appendix B)

Kevin Earley
Anchor Block Company
(Chapters 5, 8)

Bethany Eisenberg, LEED AP
VHB
(Chapters 1, 8, 10, Appendix A)

Kathryn M. Gwilym, PE, LEED AP
SvR Design Company
(Chapter 7)

Liv M. Haselbach, Ph. D., PE,
LEED AP, BD+C
Washington State University
(Appendix C)

William F. Hunt III, Ph.D., PE
North Carolina State University
(Chapter 8)

Shohreh Karimipour, PE
NYS Department of Environmental Conservation
(Chapter 9)

John T. Kevern, Ph.D., PE, LEED AP
University of Missouri-Kansas City
(Chapter 3)

Lolly Kunkler, PE
SvR Design Company
(Chapter 7)

Timothy P. Lowry, PE, LEED AP
City of Seattle
(Chapter 3)

Scott D. Perry, CPSWQ
Imbrium Systems
(Chapter 6)

Andrew Potts, PE, CPESC, LEED AP
CH2M HILL
(Chapter 2)

Ron Putz
Teufel Commercial Landscape
(Chapter 6)

Robert Roseen, Ph.D., PE, D.WRE
Geosyntec Consultants, Inc.
(Chapter 2)

David R. Smith
Interlocking Concrete Pavement Institute
(Chapters 4, 5, 6, 10)

Contributors and Reviewers

Kathlie Jeng-Bulloch, PE
City of Houston

William Lucas
Integrated Land Management

Brad Wardynski
Tetra Tech

Andrew Earles, Ph.D., PE, D.WRE
Wright Water Engineers

Doug O'Neill, LEED AP.
NRMCA

Harald von Langsdorff, Dipl.-Ing.
UNI-GROUP U.S.A

Elizabeth Fassman-Beck, Ph.D.
Stevens Institute of Technology

Amy Rowe, Ph.D.
Rutgers Cooperative Extension

Robert Goo
US EPA

Robert Traver, Ph.D., PE, D.WRE
Villanova University

Publication Design and Illustration (VHB)

Susan Lee
VHB

Sarah Toole
VHB

Sonia Tempesta
VHB

David Wong
VHB

External Review Committee

Ray Cody
US EPA Region I

David Hein, PE
Applied Research Associates, Inc.

Matthew Offenberg, PE
W.R. Grace and Company

Special Recognition

Special thanks to William F. Hunt III, Ph.D., PE, (the ASCE EWRI member responsible for the formation of the Permeable Pavement Technical Committee and first Chairman), for his continued contribution as an educator, mentor, and technical expert in the field of permeable pavement.

Preface

Permeable Pavement Initiative Background

In spite of the environmental advantages of permeable pavements in North America, they have typically remained outside the ordinary conventions of urban design and construction. Very few state and local agencies have standards or design guidelines for use by practicing engineers. There are even fewer references for more advanced information beyond generalized guidelines. Recognizing that currently available technical guidance with respect to the benefits of the technology for use as a stormwater management Best Management Practice (BMP) is limited, the Low Impact Development (LID) standing committee of the Urban Water Resources Research Council (UWRRC) of the Environmental and Water Resources Institute (EWRI) of the American Society of Civil Engineers (ASCE) formed the Permeable Pavement Technical Committee.

The LID standing committee co-sponsored technical sessions on permeable pavement technology as part of the National Conferences on LID technology held in Wilmington, NC (ASCE, 2007), Seattle, Washington (ASCE, 2008), San Francisco, CA (ASCE, 2010), and Philadelphia, PA (ASCE 2011).

The Committee is comprised of individuals from the academic and scientific communities, engineering and planning professions, the regulatory community, and technical representatives from the industry with expertise in permeable pavements. The Committee has reviewed the available data on permeable pavement technology and developed this committee report on the status of permeable pavement technology as a stormwater BMP.

The findings of the Committee presented here represent the current information at the time of publishing relative to the design, installation and maintenance of permeable pavements. This document is intended to aid the practicing engineer by presenting a brief summary of currently accepted procedures. It is not intended as a substitute for engineering experience and judgment, nor is it a replacement for more detailed standards, texts and references in the field that will be required to complete engineering projects.

The Committee recognizes that the practice of stormwater management, including the use of permeable pavements, is a dynamic and rapidly changing field with new techniques, materials, and equipment introduced continually. The Committee emphasizes that practitioners in the field must constantly be aware of new developments and modify their practice accordingly. Consequently, this document should be considered a work in progress and as the technology matures and progresses, the information presented herein will be subject to changes and updates. The UWRRC and EWRI invite comments and recommendations for improvement for possible inclusion in future updates of the document.

Disclaimer

No reference made in the Report to any specific method, product, process, data, or service constitutes or implies an endorsement, recommendation, or warranty thereof by any of the contributors. The information in the Report is for general information only and do not represent a standard of the contributors, nor is it intended as a reference in purchase specifications, contracts, regulations status, or any other legal document. The contributors make no representation or warranty of any kind, whether express or implied, concerning the accuracy, completeness, suitability, or utility of any information, apparatus, product, or process discussed in this publication, and assumes no liability therefor. This information should not be used without first securing competent advice with respect to its suitability for any general use or specific application. Anyone utilizing this information assumes all liability arising from such use, including but not limited to infringement of any patent or patents.

Introduction

Many institutions and private industry organizations have prepared guidance documents with the latest scientific and engineering information on proper design, installation, maintenance, and documented benefits of permeable pavements. However, there currently is no central clearing house for disseminating this information. The goal of this ASCE EWRI document was to review the most up-to-date information and references from academic institutions, industry groups, the engineering community, and the scientific community, and publish the results in permeable pavement in a single document.

A brief background in permeable pavements begins with pervious concrete and porous asphalt. Pervious concrete, sometimes referred to as "no-fines" concrete or "gap-graded" concrete, has been used on roadways in Europe and the United States (US) for decades. (American Concrete Institute "Pervious Concrete," NRMCA Publication MSP 68, April 2006).

Porous asphalt, "open-graded friction course" being one type of example, has also been used to enable faster road surface drainage and to reduce noise. According to the National Asphalt Pavement Association (NAPA), "in the late 1960s, the concept of porous pavement was proposed to promote percolation, reduce storm sewer loads, reduce floods, raise water tables, and replenish aquifers." The Environmental Protection Agency (EPA) contracted studies in the 1970s to "determine the capabilities of several types of porous pavements for urban runoff control, in terms of cost and efficiency." One of the earlier installations of porous asphalt was a 1977 demonstration project at the visitor parking lot at the Walden Pond State Reservation in Concord, MA, shown in **Figures I-1 and I-2**.

Figure I-1
"Pavement That Leaks" signage Walden Pond, Concord, MA
Source: A. Richard Miller (http://millermicro.com/porpave.html)

Figure I-2
Porous asphalt built in 1977 Walden Pond, Concord, MA
Source: A. Richard Miller (http://millermicro.com/porpave.html)

Permeable pavement technology is continuing to evolve as the goals for managing stormwater runoff increase. The uses, installation methods and types of permeable pavement materials are also rapidly expanding. NAPA reports and other studies point out that while there have been failures in some pavements, these are most commonly attributed to poor site selection, installation, and/or material specifications. Loss of permeability for pavements is typically attributed to fine materials that are being allowed to run-on and collect on the surface, causing clogging. While well designed, installed, and maintained pavements are proven to have high benefits without failure, past failures have resulted in a lack of confidence in the widespread use of permeable pavements in many areas.

Given the evolution in permeable pavement technology and studies, addressing the need for better designs, construction, and maintenance is the goal of this report. This report is organized into ten chapters with appendices. **Chapter 1** presents an overview of permeable pavement systems and describes design considerations that apply to most permeable pavement types and applications. The chapter concludes with useful checklists on design and construction. Currently available design, use, and performance information for standard permeable pavement types including porous asphalt, pervious concrete, permeable interlocking concrete pavement, and grid pavements are provided in greater detail in **Chapters 2 through 5**. Each of these chapters starts with a fact sheet summarizing the uses, benefits, and limitations of each permeable pavement type. **Chapter 6** describes other types of permeable pavement materials and applications not typical to standard permeable pavement. **Chapter 7** provides insights and experience-based recommendations into aspects that contribute to successful permeable pavement projects. **Chapter 8** covers maintenance common to all standard permeable pavements. **Chapter 9** provides an overview of hydrologic design models. These include sizing models for static design and time-based simulation models. **Chapter 10** identifies key areas where additional research is recommended. These include full-scale validation of hydrologic models and structural design approaches; specific design details for system options and various locations; continued research on pollutant removal capabilities of the various pavements; and specifications and guidelines for improved construction and maintenance practices.

The appendices provide additional reference information. **Appendix A** contains a simple fact sheet addressing permeable pavement concerns such as clogging, performance, use in cold climates, and maintenance requirements. Design, cost, and performance summary tables for each of the four typical permeable pavement types are included in **Appendix B**. **Appendix C** provides helpful commentary on standards, specifications, testing methods, resources, and references on various permeable pavement systems. Finally, **Appendix D** presents a glossary of terms.

1 Design Considerations Common To All Permeable Pavements

1.1 Introduction

Impervious cover in watersheds without controls result in increased stormwater runoff and decreased groundwater recharge in response to rainfall events. This increased runoff can be the cause of negative impacts to municipal drainage systems, natural resource areas, and private properties. Impacts from impervious cover runoff can include increased stormwater flow volumes and velocities that can cause flooding and/or erosion; the transport of stormwater-related pollutants/debris from impervious surfaces to resource areas and drainage system components; the reduction of rainwater recharge to support groundwater systems and stream baseflow; and overall physical and chemical changes to the natural hydrologic system in a watershed. A key goal for improving and/or protecting the nation's water resources is the reduction of impervious cover in watersheds. For more than 25 years in the United States, there has been a continual increase in the development and implementation of stormwater management regulations with new requirements and design guidelines. These efforts are focused on preventing negative impacts as described above. Permeable pavement is a stormwater management practice that can reduce impervious surfaces in watersheds and related impacts.

Design professionals, scientific researchers, and supporting industries have advanced the use of permeable pavements with new and refined construction methodologies, testing standards, and materials for many pavement types. The development of additional resources to support the proper design, installation, and maintenance of permeable pavements and a mechanism to promote standardization of practices and sharing of information continues to be needed. This document is intended to disseminate currently available technical information to practitioners, identify needs for additional research and technical assistance, and provide a resource for individuals interested in the design and use of permeable pavements.

Permeable Pavement Systems as a Stormwater Management Best Practice

Current practice documents and design guidelines that require or promote the implementation of good stormwater management practices often refer to these recommended practices as "Best Management Practices" (BMPs) or "Stormwater Control Measures" (SCMs). These BMPs are typically grouped into two categories: structural BMPs and non-structural BMPs. Structural BMPs are practices where something is actually built or "constructed" to manage the quantity and/or quality of the stormwater runoff. Non-structural practices are usually methods to prevent increased runoff or control the quality of the runoff without constructing something. An example of a non-structural practice is the development of a law or voluntary practice guideline to eliminate the use of phosphorus (P) containing fertilizer in a watershed, reducing P pollutant loads carried in stormwater runoff to water resources.

Permeable pavement systems, consisting of a surface layer through which water passes underlain by permeable base/subbase materials, are a structural stormwater management practice. This practice results in runoff reduction (or runoff elimination with proper site conditions), water quality treatment of stormwater runoff as it infiltrates through the pavement surface, base/subbase materials, and the application of a pavement system that can support appropriate goals/needs for pavement usage in a built environment while more closely mimicking the natural hydrologic responses to rainfall events. **Figure 1-1** depicts a concrete grid type permeable pavement adjacent to a heavily used paved pathway in an urban setting.

Figure 1-1
Castellated concrete grid pavement at walkway edges, Cambridge, MA
Source: Hastings Pavement Company

Benefits of Permeable Pavement Systems

Key benefits of using permeable pavement systems instead of impermeable pavement include:

1. **Reduces runoff volumes and peak discharge rates**
   - Reduces total storm runoff volumes and spreads the volume over a longer time period compared to impervious pavement, helping reduce flooding impacts
   - Reduces volumes discharging to municipal storm drainage systems and conveyance channels, helping reduce combined sewer overflows and conveyance system impacts
   - Helps developed sites more closely match pre-development hydrology in terms of runoff volume and peak rates of discharge
   - Reduces pollutant load contributions as runoff volume reduction is directly related to pollutant load reductions
   - Delays and reduces peak discharge rates (for partial-infiltration and no-infiltration system designs)

2. **Increases infiltration/recharge (if soils allow and no liner is required)**
   - Increases groundwater recharge, supporting natural stream baseflow and drinking water supplies
   - Infiltration/recharge may result in more available water for nearby vegetation/trees
   - Infiltration/filtration through base/subbase and/or uncompacted soil subgrade provides water quality treatment

3. **Improves water quality**
   - Pavement and/or base/subbase materials contribute to the removal of heavy metals, oils/grease, and nutrient loads as stormwater filters through the system (amounts vary with design)
   - Helps reduce or eliminate deposition of sand on pavement and contributions in runoff that would result if sand was used for winter conditions; this also eliminates the maintenance requirements for cleaning of deposited sand on pavement and in catch basins or other stormwater management practices
   - Eliminates or reduces the use of salt/deicing chemicals for winter pavement management

4. **Reduces stormwater temperature and heat island effects from pavement**
   - Cooler stormwater runoff can result from the light color of some pavement surfaces as well as by infiltration through bases/subbases
   - Stormwater is cooled as it filters through the base/subbase

5. **Better site design**
   - Reduces or eliminates the need for detention/retention ponds and related safety and/or mosquito breeding concerns
   - Can result in preservation of the natural landscape (i.e., woods and open space) that might have been replaced by detention/retention ponds
   - Promotes tree survival and vegetated growth near paved areas by providing air and water in the root zone

6. **Reduces drainage system infrastructure and costs**
   - Reduces or eliminates the need for catchbasins, manholes, and storm drains for a piped drainage system resulting in cost savings (initial, operation, and maintenance costs)
   - Can result in a reduction or elimination of a stormwater utility fee

7. **Pavement surface benefits**
   - Reduces hydroplaning risks on roadways and standing water on pedestrian walkways
   - Reduces wet-weather glare
   - Reduces heat island effects due to voids that allow for more air movement through pavement (and cooler surface temperatures if a light colored pavement surface is used)
   - Lowers risk of damage due to freeze/thaw conditions such as heaving and crack formation compared to conventional pavements
   - Reduces or eliminates thin ice accumulations on pavement due the melting/re-freezing cycles during sunny/warmer days with colder nights in the winter season
   - Increases vehicle safety with the benefits of the surface pavement as listed above

Figure 1-2
Standard concrete pavement (left) next to pervious concrete (right) during winter, Iowa State University, Ames, IA
Source: John Kevern

Figure 1-2 shows standard concrete pavement during winter with snow melt and ice accumulation adjacent to a permeable concrete parking section.

Applications for Permeable Pavement

Permeable pavements are commonly used for walkways, driveways, patios, courtyards, sidewalks, parking lots, alleys, and in low-volume roadways, generally with posted speed limits of 55 kph (35 mph) or lower. Permeable pavements are also used in recreational and park-related applications such as playground spray pools, areas around water fountains, or as permeable buffers around tree beds and planters. Many applications have been used to support outdoor uses that require or benefit from water/stormwater infiltration from paved surfaces as opposed to ponding and/or runoff, such as entryways to eliminate ponding at doors.

Since permeable pavements can retain vehicular and pedestrian functionality, they are highly appropriate for use in geographically-constrained urban areas where the dual purpose of stormwater management and a usable surface is provided. Further, permeable pavements can be used in retrofit applications to provide stormwater management in space limited locations. Permeable pavements can be strategically placed to accept clean run-on from adjacent uses such as walkways or roofs. As described in this document, the type of permeable pavement material selected and the design of the full pavement system will be dependent on the site goals and the particular use for the pavement.

Permeable pavements are currently not recommended for use with consistent heavy loads/truck traffic, high speed/high volume roadways, or areas that may have a higher than typical potential for pollutant, sediment deposition, or organic matter accumulation. Pavements that may experience repetitive turning motions in the same location are also not good candidates for many permeable pavement surfaces, especially porous asphalt. The specific product selected for this use should be evaluated.

The following sections describe permeable pavement systems and design considerations applicable to all pavement applications.

1.2   Permeable Pavement Systems

As shown in **Figure 1-3**, a typical permeable pavement system cross-section includes a permeable pavement surface layer on top of open-graded aggregate base/subbase layers.

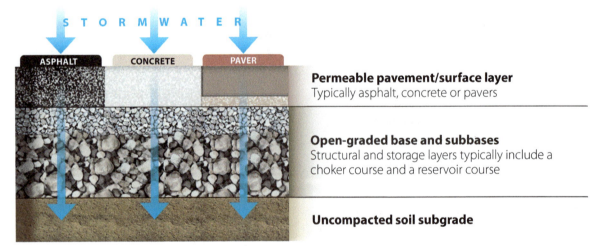

Figure 1-3
Generic permeable pavement cross-section
Source: © VHB

As stated in the Virginia Department of Conservation and Recreation (DCR) (2011) *Stormwater Design Specification No. 7 for Permeable Pavement*:

> "*Permeable pavements are alternative paving surfaces that allow stormwater runoff to filter through voids in the pavement surface into an underlying stone reservoir, where it is temporarily stored and/or infiltrated. A variety of permeable pavement surfaces are available, including pervious concrete, porous asphalt and permeable interlocking concrete pavers. While the specific design may vary, all permeable pavements have a similar structure, consisting of a surface pavement layer, and an underlying stone aggregate/reservoir layer. The thickness of the reservoir layer is determined by both a structural and hydrologic design analysis. The reservoir layer serves to retain stormwater and also supports the design traffic loads for the pavement. If infiltration rates in the native soils permit, permeable pavement can be designed without an underdrain, to enable full infiltration of runoff.*"

Permeable Pavement Types

Permeable pavements are typically categorized into four major categories as described in **Table 1-1** (modified from Ohio Stormwater Manual 2006 and NCDENR Stormwater BMP Manual 2012).

Table 1-1 Typical Permeable Pavement Types

MATERIAL AND DESCRIPTION	DETAIL
Porous asphalt (PA) Porous asphalt is similar to conventional asphalt, except the fines are removed to create greater void space. Additives and higher-grade binders are typically used to provide greater durability and prevent draindown of the asphalt binder.	
Pervious concrete (PC) Pervious concrete is produced by adding aggregate into a cementious mix to maintain interconnected void space. As a result, it has a coarser appearance than standard concrete. Additives may be combined to increase strength and improve binding.	
Permeable interlocking concrete pavement (PICP) Permeable interlocking concrete pavement is made of interlocking concrete pavers that maintain drainage through stone-filled gaps between the pavers. The pavers are not permeable.	
Grid pavement systems (plastic or concrete) Grid pavement systems are modular grids filled with turf and/or gravel. Open-celled concrete or plastic structural units are typically filled with small uniformly graded gravel that allows infiltration through the surface.	

Design, use, and performance information for these standard permeable pavement types are provided in Chapters 2 through 5. Chapter 6 describes other types of permeable pavements (i.e., permeable pavers and rubber overlays) with varied materials and applications, not included in this list of typical pavement types. Permeable pavers pavement differ from PICP because the pavers themselves are permeable. Design, cost, and performance summary tables for each of the four typical permeable pavement types are included in Appendix B.

In general, the most significant difference between asphalt and concrete-based permeable pavements versus standard impermeable pavement is the removal of the fines in the mix, which results in the creation of void space that makes the pavement permeable. The removal of the fines in the pavement mix changes the structural qualities of the pavement surface and affects the durability of the pavement as well as the physical qualities of the mix. These changes can result in the need for modifications to field handling, installation techniques, and equipment. For porous asphalt, additives are often combined with the pavement mix to increase pavement durability as well as prevent draindown and sealing of the asphalt binder. For pervious concrete pavements, additives may be included to promote initial hydration and increase strength development.

For all permeable pavements, the structural strength of the entire pavement system can be increased with an increase in thickness and modification of materials used for the base/subbase aggregates. The design engineer must select the appropriate pavement surface type, base/subbase materials, and depth of materials to support the structural load requirements for the selected use.

Permeable Pavement System Designs

In addition to the structural design, the engineer must design the pavement system to achieve stormwater management goals, while also considering site-specific characteristics. Designers should consult with regulatory requirements as described in Section 1.3 Regulatory Requirements. Stormwater management goals may include designing for the removal of a targeted pollutant or for a specific reduction in volume and/or peak flow rate of stormwater (design storm). Characteristics that drive the site design generally include site hydrology, soil subgrade permeability, groundwater depth, subsurface constraints (i.e., wells, septic systems), and project budgets. These factors, as described herein, will determine the appropriate permeable pavement type and the structural/materials design required for each application. The designer would then determine the details for the pavement system cross-section, as generically shown in Figure 1-4. For many permeable pavement applications, this base/subbase would typically include a choker course beneath the pavement surface and a reservoir course below the choker course, underlain by the uncompacted soil subgrade.

Figure 1-4 also shows a perforated underdrain pipe, which may be required, towards the base of the reservoir course. Underdrain designs are common components for many pavement systems when subsurface and/or site conditions limit the volume or prohibit the recharge of stormwater into the soil. This may be a result of low-permeable natural soils, high groundwater elevations, contaminated soils prohibiting infiltration practices or shallow bedrock. Underdrains can be placed at the bottom of the pavement base/subbases in the reservoir course, as shown in Figure 1-4. Specific design modifications such as weirs or upturned elbows may be included by engineers to avoid rapid dewaterting, increase temporary storage, and promote infiltration by ponding water in the reservoir.

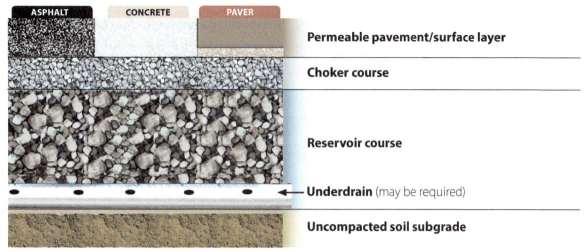

Figure 1-4
Typical permeable pavement base/subbase layers and underdrain design
Source: © VHB

Reservoir/Infiltration Storage

Permeable pavement systems can be designed to infiltrate all-stormwater (full-infiltration), some-stormwater (partial-infiltration), or no-stormwater (no-infiltration) from the aggregate reservoir layer into the soil subgrade. Full- or partial-infiltration designs with recharge to underlying soils provide a natural pathway for managing stormwater rather than the direct discharge of stormwater to piped drainage systems. Examples of each infiltration type are illustrated in **Figures 1-5 through 1-8**.

Full-infiltration designs generally do not use underdrains and are typically in areas with high permeability native sandy soils. **Figure 1-5** shows a typical full infiltration system cross-section.

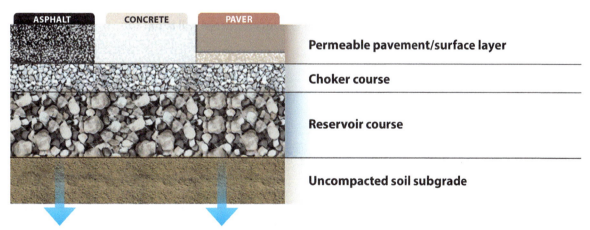

Figure 1-5
Full-infiltration design—No underdrain
Source: © VHB

Partial-infiltration designs infiltrate some water into the soil subgrade, typically into lower permeability soils, with the remaining filtered water exiting via the perforated underdrain (**Figure 1-6**). Local regulations often require draining of a subsurface storage system within 48 to 72 hours after the end of the rainfall event.

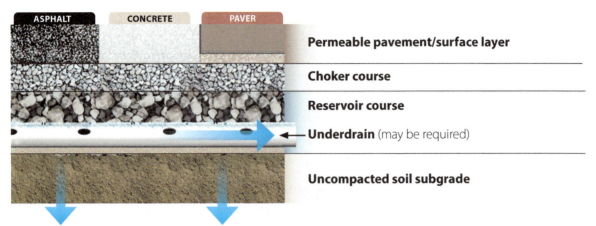

Figure 1-6
Partial-infiltration design—Underdrain, no liner
Source: © VHB

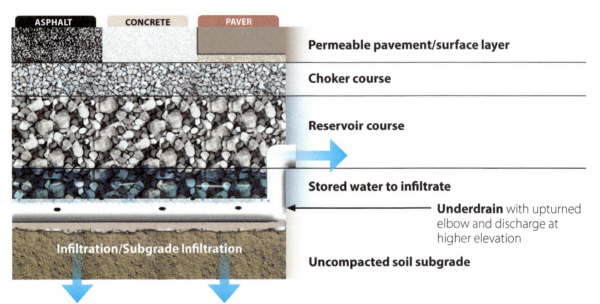

Figure 1-7
Partial-infiltration with inverted underdrain
Source: © VHB

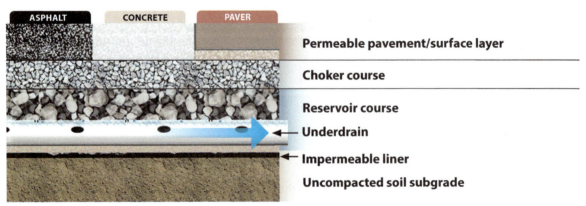

Figure 1-8
No-infiltration design—Underdrain and impermeable liner
Source: © VHB

A raised or "upturned elbow" underdrain (**Figure 1-7**), or other flow restriction device, is often included in the reservoir course to create a hydraulic head to promote greater infiltration and/or increase the detention/storage capabilities of the system. These designs are often similar to subsurface detention basin system designs and may include an outlet control structure that handles both low-flow discharges and overflow bypasses.

No-infiltration systems are designed with an impermeable liner to prevent any infiltration of stormwater into the existing subgrade (**Figure 1-8**). The liner could be either a geosynthetic material or an impermeable barrier such as clay. This system requires storage in the reservoir course and release via the underdrains.

Permeable pavement systems with underdrains, even if connected to a piped stormdrain system, provide a delay in runoff volumes as well as water quality treatment. Increased volume reductions may also be realized if the underdrain is not at the bottom of the reservoir and/or if it discharges to another stormwater management BMP with storage and/or infiltration capabilities.

Overflow Conveyance

The design of all permeable pavement systems must include a method for safely conveying overflows of the system. There are two situations that can result in overflow: (1) the rain storm is so intense that not all water can infiltrate the surface of the permeable pavement and will create some surface runoff, and (2) overflow of the storage layer occurs. For partial- and no-infiltration systems, it is critical to design the overflow to handle the flows beyond the design storm capacity. While full-infiltration systems are designed to infiltrate entire rainfall volumes, a safe overflow conveyance or bypass for high-volume and intensity storm flows that may exceed the capacity of the system must be included in the design. The overflows can be directed to surface conveyance channels and/or a closed drainage system. Overflow/bypass designs may include underdrains, manholes, catchbasins, curb cuts and other structural components in various configurations. Some examples include:

- Underdrains in the reservoir course connected to a manhole structure with a weir or other controlled outlet design
- A catch basin inlet at the pavement surface with an underdain connected from the reservoir course to the catch basin
- A surface flow bypass via a curb cut to a stabilized channel or other conveyance structure

The NCDENR (2012) *Stormwater BMP Manual Section 18* on permeable pavement includes detailed drawings of a variety of configurations for outlet control and bypass control structures.

Filter Courses

While not a typical design, permeable pavement systems can be designed with filter courses in the base/subbase layer to provide additional water quality treatment (**Figure 1-9**). The depth and materials used for filtering are designed according to the physical flow requirements and the targeted contaminant(s) to be removed. Typically, a smaller stone (e.g., choker course) layer is provided below the filter course to prevent the filter media from moving into the reservoir course below. Regardless of the presence of a filter layer, the process of runoff flowing through the pavement base/subbase layers provides water quality benefits.

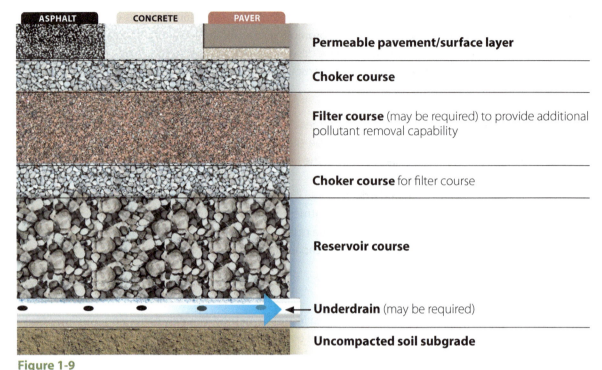

Figure 1-9
Typical permeable pavement with filter course for additional water quality treatment
Source: © VHB

Typical Permeable Pavement System Materials

As shown in **Figure 1-3**, permeable pavement systems generally have a pavement surface with various base/subbase courses above the existing uncompacted soil subgrade. The following section describes the typical components of a permeable pavement system. The material specifications for base/subbases differ with the type of permeable pavement surface used. In all cases, a site-specific design load analysis for the intended pavement use should be conducted by a qualified engineer to determine the exact design criteria for the base courses. The following descriptions are general and must be modified for the specific use.

Typical materials used beneath the permeable pavement surface include the following:

Bedding Course

Permeable interlocking concrete pavement (PICP), permeable pavers, and concrete and plastic grid pavements typically include a small-sized aggregate bedding layer below the pavement surface and on top of the choker course to ensure a level surface for the pavers/grids. This is not required for pervious concrete (PC)or porous asphalt (PA) applications.

- PA—Not required
- PC—Not required
- PICP—ASTM No. 8 stone or per manufacturers specifications
- Pervious pavers and grids—Bedding course per manufacturers specifications

Choker Course

This aggregate stone layer is used to level out the top of the larger stone aggregate subbase/reservoir course to provide a smooth level surface for the pavement surface. The typical system design includes a choker course layer beneath the permeable pavement surface or the bedding course (as applicable) as follows:

- PA—ASTM No. 57 stone over reservoir course (optional).
- PC—ASTM No. 57 stone over reservoir course (optional)
- PICP—ASTM No. 57 stone
- Pervious pavers and grids—ASTM No. 57 stone.

Reservoir Course

The reservoir course is a layer of clean, washed stone designed with a depth to support the structural load requirements as well as water detention/retention requirements. The thickness of the reservoir course and the need for an underdrain depends on the permeability of the existing soil subgrade, structural requirements of the pavement subase, depth-to-water table/bedrock, or other confining layer and potential frost depths (Virgina DCR 2011).

The reservoir course for a typical system design would be laid over the existing uncompacted soil subgrade with common stone sizes as follows:

- PA—ASTM No. 2 or No. 3 stone
- PC—ASTM No. 2, No. 3 or No. 57 stone
- PICP and pavers—ASTM No. 2 or No. 3 stone
- Grid pavement—ASTM No. 2 or 3 stone

Underdrains

Underdrains are commonly included within the subbase of permeable pavement system designs for a variety of reasons including:

- Impermeable/Very low permeability of existing subgrade soils
- Lowered permeability of subgrade soils due to compaction
- Placement of an impermeable barrier (i.e., liner) to prevent infiltration into existing subgrade soils (e.g., due to contaminated soils)
- Close proximity to subsurface features (e.g., utility trenches, basements, wells, septic systems) that would be negatively impacted with full- or partial-infiltration of design storms
- Proximity to an exposed slope
- Unacceptable separation from groundwater, bedrock, or other confining layer

Designs with underdrains typically include a depth of aggregate above and below the drains. The drains are typically 10 to 15 cm (4 to 6 in.) diameter perforated PVC pipes. The upgradient end of the underdrain should be capped. There should be no perforations within 0.3 m (1 ft) of the structure where an underdrain pipe is connected to a structure (Virginia DCR 2011).

Geosynthetic Materials

As described in the NCDENR (2012) *Stormwater BMP Manual*, permeable pavement systems may include geosynthetic materials such as geotextiles, geomembranes, and geogrids. The guidance of an experienced civil and/or geotechnical engineer familiar with local site soils conditions, pavement design, and stormwater management should be sought to confirm the suitability of the soil characteristics and possible use of geosynthetics in a permeable pavement system. These supplemental materials are described here:

Geotextiles

Geotextiles are typically placed vertically against excavated walls (**Figure 1-10**).

> *"Geotextiles are permeable geosynthetics that prevent the intrusion of native soils into the aggregate subbase. If no full-depth curb or impermeable barrier separates the permeable pavement from the adjacent soil, full depth geotextile is recommended to vertically separate the native soil from the aggregate with a minimum 0.3 m (1 ft) of the geotextile lying horizontally on the soil subgrade (NCDENR 2012)."*

Upper layer, horizontally placed geotextiles are generally not recommended for use above the open-graded-base material of a permeable pavement system, based on research by Rowe (2010). The use of geotextiles at the bottom of the reservoir may be used if warranted, based on an evaluation by the geotechnical engineer. If soil stability is a concern, the geotechnical engineer should design the appropriate treatment and may consider a stabilizing geotextile, often called a "geogrid."

Figure 1-10
Graphic of vertical geotextile
Source: © VHB

Geogrid

A geogrid is a polymer, net-like structure that can be used to reinforce and confine materials as well as distribute loads over weak soil subgrades (**Figure 1-11**).

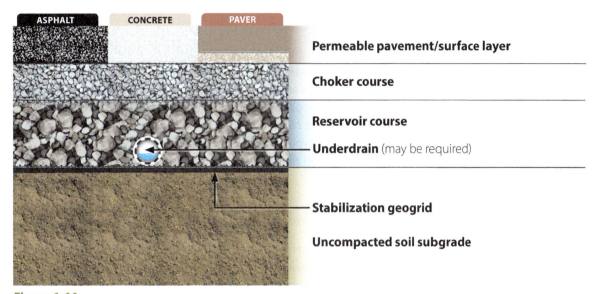

Figure 1-11
Graphic of geogrid placement for soil stabilization
Source: © VHB

Geomembrane

A geomembrane is an impermeable geosynthetic used to act as a barrier to the movement of water or gas. These are often referred to as impermeable "liners" and are used when the movement of stormwater into the existing soil subgrade is not desired. Some manufacturers' specifications require or recommend that a layer of sand be placed beneath the geomembrane to prevent tears or punctures when the aggregate rock is placed on top of it (**Figure 1-12**).

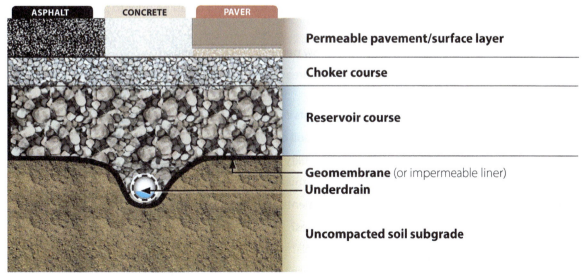

Figure 1-12
Graphic of geomembrane
Source: © VHB

Manufactured Underground Storage Systems

While most permeable pavement installations are underlain with a stone aggregate bed for water storage (reservoir course), alternative manufactured storage products can offer additional capacity. These may be placed within the subbase layers and include plastic storage structures and precast concrete storage units.

Edge Restraints

Permeable pavements typically require perimeter edge restraints or special joint methods when used adjacent to impermeable pavements. Edge restraints are needed to prevent raveling and breakdown of the pavement edge. Flexible, plastic, or metal edging supported with spikes is not recommended for pavements supporting vehicular use. Restraints can be concrete curbing flush with the pavement or other rigid edge control materials recommended for the pavement type and use. Some designs include the edge at a slightly higher elevation to help keep stormwater on the permeable surface and/or prevent fines from traveling to the pavement.

Intersection of Permeable and Impermeable Pavements

A variety of design approaches and materials are used for the intersection between permeable pavement and conventional pavement. It is important that the sides of the reservoir course be lined with an impermeable liner or barrier to prevent stormwater from entering the subbase of the conventional pavement. It is recommended that a detail be prepared to specify the construction process at the pavement intersection between two pavement types. For maintenance purposes, a visual delineation between pavement types may be beneficial.

1.3   Regulatory Requirements

Applicable federal, state, and local regulations must be reviewed prior to initiating a permeable pavement design. While some regulations promote or may require the use of permeable pavement, there may be certain restrictions, specific design requirements, and potentially a prohibition of the use of permeable pavements for certain site conditions. Additionally, there may also be some benefits or incentives offered for using permeable pavement. Some essential questions that should be asked prior to proposing permeable pavement use include the following:

- Do the applicable local regulatory agencies allow permeable pavement use?
- Are permeable pavements prohibited in certain areas such as groundwater recharge zones or areas with high/very high infiltration rates?
- Are there prohibitions or specific design requirements if permeable pavement is used for roadway applications?
- Are permeable pavements prohibited in certain soil conditions such as low permeability or fill?
- Are there credits offered to reduce stormwater utility fees or permitting fees?
- Are site development credits offered by using permeable pavement?
- Are there regulatory hydrologic control requirements?
  - Are they for volume and/or peak discharge rates?
  - What is the design storm for hydrologic controls?
- Are there water quality control requirements specific to using permeable pavements?
- Are there specific pollutants of concern that need to be addressed in terms of water quality treatment?

- Are there specific design guidelines or specifications mandated under applicable federal, state, or local regulations?
- How do Underground Injection Control (UIC) regulations apply to permeable pavement systems?

These questions and others related to the design process are included in the **Checklist 1: Design Considerations Common To All Permeable Pavements Summary** provided at the end of this chapter. Answers to these preliminary design-related questions will help designers determine whether a permeable pavement system is an appropriate stormwater management/site development practice for a particular site or location and may drive the preliminary sizing and selection of materials of the system. If the goal of the system design is for the water quality management of smaller and more frequent storm flows (first flush—i.e., larger flows managed with other on-site stormwater system features and traffic loads are expected to be low and light duty), the reservoir depth of a permeable pavement system would be much smaller than if designed more for flood control purposes.

While regulatory requirements set the stage for design goals and requirements, the regional conditions, site specific conditions, and intended use of the pavement will drive the actual design as described in the following sections.

1.4 Regional Considerations

While engineering details and specifications are available for certain permeable pavements, specific practices and materials included in existing specifications may not be transferable from one region to another. Two key regional factors that affect permeable pavement design are climate and locally available materials (including material type and distance of supplier from project site).

Conditions such as temperature, precipitation type, and patterns, as well as the availability of permeable pavement system materials may require modifications to more typical specifications currently being used in certain locations across the country. These modifications are critical to ensure the appropriate selection of materials for the climate and use, and may affect the cost. The differing costs may result in a decision to not choose a specific type of pavement. Also, certain types of pavements are not suitable for use in some climates. For example, porous asphalt is not used in areas with high year-round high temperatures. The design of the pavement with the absence of fines in the asphalt pavement mix can result in the softening of porous asphalt pavements and the draindown of the binder into the lower layers. This results in pavement degradation and reduction or elimination of permeability as the binder collects in the lower layers of the subbase. In addition, particles may embed within the binder and further decrease infiltration.

If regional specifications for the selected permeable pavement are not readily available, the design engineer should work closely with pavement materials suppliers to prepare specifications consistent with pavement load and hydraulic function requirements.

1.5 Site Conditions

A review of site conditions is necessary to determine the permeable pavement design options and preliminary features, such as:

- The preferred location on the site
- The base/subbase design that will support the desired structural and stormwater management goals
- Whether or not an underdrain system or impermeable liner with underdrain system is required.

Soils Conditions/Infiltration

Permeable pavements can be used successfully in most soil types, but the particular soil characteristics will influence how the subgrade is prepared, the particular dimensions of the pavement reservoir and the use of underdrains.

As described in Section 1.2 Permeable Pavement Systems, permeable pavements can be designed to infiltrate all, some, or none of the stormwater from the aggregate storage layer into the soil subgrade. The existing soils and site conditions drive the need or desire for underdrains. Full- or partial-infiltration designs with recharge to underlying soils provide a natural pathway for managing stormwater as compared to discharges to a closed pipe drainage system.

No-infiltration systems are designs that include an impermeable barrier on the bottom and sides of the subbase and underdrains, thus preventing the movement of stormwater into adjacent and underlying soils. Water drains from the reservoir into perforated underdrains. This design is typically used in marginal or contaminated soils to prevent infiltration into subgrade soil. It is recommended that no-infiltration designs be used if the following conditions are present:

- The distance between the permeable pavement and water supply wells is less than 30 m (100 ft). Designers should consult local regulatory agencies for additional guidance or varied regulations.
- The soil depth-to-groundwater is insufficient to offer adequate filtering and treatment of stormwater related pollutants. Local regulations often require a minimum flow rate through a minimum depth of soil. This will vary given the use of the pavement, depth of subbase layers, soil permeability, and depth-to-groundwater. A minimum 0.61 m (2 ft) separation from groundwater is recommended.
- The system is directly over solid rock or impermeable rock/soil layer such as compacted glacial till with no loose permeable rock layer above it.
- The system is near drinking-water aquifers without the minimum 2-foot vertical separation or sufficient soil permeability rates to filter pollutants before they enter the groundwater.
- The system is over some fill soils that have unacceptable stability when exposed to infiltrating water such as expansive soils or poorly compacted fill soils.
- The pavement is adjacent to fill or natural slopes where soil conditions may result in lateral breakout of the stormwater on the slope (a lateral impermeable barrier may overcome this situation and allow the design of a full- or partial-infiltrating system).
- The location is in an area where stormwater may be exposed to hazardous materials as a result of land use or the potential for an accidental spill of hazardous materials is higher than normal (i.e., "stormwater hotspot"). Some examples include fueling stations and salvage yards.
- The location is in an area with karst geology with limestone deposits subject to sinkhole development due to underground artesian water movement. A geotechnical engineer is required in these areas as some sites may not be compatible with any permeable pavement.
- The pavement systems near building foundations and basements are subject to flooding. They are not recommended for use if within 3 m (10 ft) unless adequate perimeter drainage, waterproofing, and geotechnical designs are completed and approved.

While there is no significant stormwater volume reductions realized with no-infiltration designs, they can be designed to control discharge rates and offer some water quality treatment. Volume reductions are realized if the underdrain discharges to another infiltrating stormwater management BMP.

Clay and Low Permeability Soils

Permeable pavement systems can be used above clay and low permeability soils with an underdrain. Designs on poor soils often include underdrains placed at a specifically determined height above the bottom of the system to allow for ponding in the reservoir layer and increased infiltration capacity into the subgrade soils. **Figure 1-8**, previously described, shows an elbowed perforated underdrain within the pavement reservoir layer to encourage downward and lateral discharge. These systems have the potential to provide greater infiltration into many types of lower permeability soil groups classified by the Natural Resources Conservation Service (NRCS) as hydrologic soil group (HSG) C and D soils. Exceptions are noted for soil types with high shrink/swell potential, as described in the next section.

Shrink/Swell Soils

Permeable pavement systems designed to infiltrate into the soil subgrade should not be sited on soils with high shrink/swell potential due to a high risk of subsurface and surface movement. Systems with impermeable liners are an option for this application. A geotechnical engineer should be involved to determine the suitability of the site for permeable pavements, the design of the geomembrane (liner), and/or use of other geotextiles or materials for the entire pavement system stability.

Pretreatment

As stated in the Virginia DCR (2011) *Stormwater Design Specification No. 7*:

"Pretreatment for most permeable pavement applications is not necessary, since the surface acts as a pretreatment to the reservoir layer below. Additional pretreatment may be appropriate if the pavement receives run-on from an adjacent impervious area."

For example, a gravel filter strip or other filtering device could be provided between the impermeable and permeable pavements to trap sediments before they reach the permeable pavement.

Groundwater

In most cases, the bottom of a permeable pavement system should not intersect with groundwater. The separation depth from the bottom of the permeable pavement to groundwater should:

- Satisfy local/state infiltration system/groundwater separation requirements, if applicable. Generally, a minimum 0.6 m (2 ft) separation distance from seasonal high groundwater is recommended.
- Allow for proper draining of the reservoir course to ensure that storage is available for frequent design storms and that all the the subsurface aggregate bases will drain properly, especially in cold weather climates. Local jurisdictions may specify a time for reservoir draindown.
- Be sufficient to prevent unacceptable groundwater mounding conditions.

All requirements/prohibitions should be reviewed if the site is in a water resources protection area, groundwater, drinking water supply, recharge zone, or wellhead protection zone.

Slope/Contours

Existing and proposed slopes of the project site should be evaluated when considering a permeable pavement system.

Adjacent Lateral Slopes

Lateral impermeable barriers may be necessary if the subsurface reservoir/infiltration area is designed in close proximity to a sloped area.

Sloped Pavement

The design of permeable pavements on slopes requires special considerations to prevent drainage of the reservoir onto the downstream pavement surface, with potential damage to the pavement and the loss of infiltration. If the top of the subgrade is sloped, engineers should consider incorporating subsurface terracing, check-dams, baffles, partitions, berms, or some combination to inhibit lateral flow and promote vertical infiltration. These practices promote infiltration across the entire area of the subgrade and reduce the potential for lateral flow out onto the pavement surface. Hydrology and soil conditions at each site are unique. When making this determination, engineers should consider the subgrade slope, subgrade infiltration rate, storage volume, site orientation, and design or performance requirements. While other guidance documents commonly provide recommendations, such as slope thresholds of 1% to 5% with soil infiltration thresholds of up to 100 mm/hr (4 in./hr), more research is needed to determine critical slope and infiltration rate thresholds that require incorporating lateral flow inhibiting structures into permeable pavement systems.

Terracing is the practice of constructing the subgrade layers in level steps or infiltration "beds." All bed bottoms should be as level as feasible to promote uniform infiltration. This requires varying cross-section thickness along the slope as shown in **Figure 1-13**. The use of sloped beds may be infeasible when constructed on steep slopes.

Check dams and barriers may be constructed as soil berms or as berms constructed by placing impermeable fabric placed around the subbase aggregates, or fabricated baffles periodically placed along the contour line to create internal grade control. Sloped systems can be designed with underdrains at the upslope side of the berms if full infiltration is not expected in each storage cell (**Figure 1-13**). A geotechnical engineer

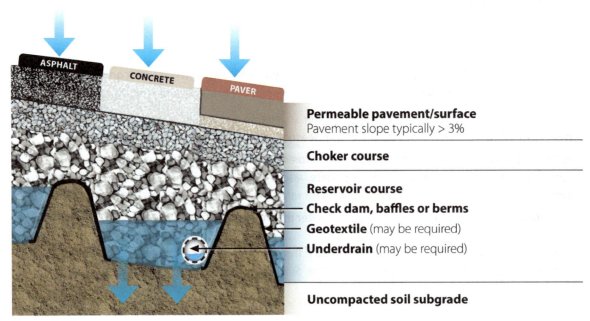

Figure 1-13
Sloped permeable pavement with checkdams, baffles or berms *(exaggerated depiction)*
Source: © VHB

should be involved with any pavement system with a sloped design or one that is near an existing or proposed slope on the site.

Surface Run-on

Permeable pavement systems may be designed to accept runoff free of sediments and contaminants from adjacent impervious and stabilized pervious areas. It is recommended to prohibit or minimize the amount of runoff from pervious areas (vegetated/landscaped) since they may contribute organic and inorganic materials and fines, which could increase the maintenance cleaning effort.

"Run-on" refers to surface flows discharged onto the permeable pavement surface. Permeable pavement systems may also be designed to accept runoff free of sediments directly into the reservoir layer depending on the systems' hydraulic capacity. A common design is to have roof drains discharge directly to the reservoir course of the pavement system or to an underground flow spreader and then into the reservoir course.

In general, the ratio of the total drainage area to permeable pavement surface area is 2:5, if there is surface run-on from impervious surfaces. However, caution should be used if there is reason to believe that run-on could transport fines or sediment to the permeable pavement surface, or the runoff is concentrated to one area of the pavement. Specifications individual to the material type apply and local code standards should be followed regarding the ratio of allowable runoff from adjacent roofs, pavements, and impervious areas to the permeable pavement surface. This is only a guideline. Site-specific conditions such as slopes, vegetation, erosion potential, and anticipated sediment loading could lower the ratio, while direct roof drain runoff piped to base/subbase could increase the ratio.

Run-on or direct discharge to base/subbase reservoir layers from areas with active construction should never be allowed, and all permeable pavements should be protected during construction. This is discussed in detail in Chapter 8.

Land Use and Clogging Potential

Existing, proposed, and neighboring land uses may create an increased risk for pavement surface clogging. If nearby soils are disturbed or unstable, the potential exists for transport of clogging materials to the pavement surface. Such threats may include:

- Sediment from adjacent soils
- Sediment from nearby construction activities that may enter the site via tires/vehicles
- Mulch from adjacent landscaping
- Sediment and debris from snow pile storage and snow melt
- Blown-on fines or materials such as sand near beach areas
- Leaf and foliage debris during autumn months

Underground Structures and Utilities

Increased infiltration through the pavement base/subbase should not impact existing underground structures, such as basements, septic systems, or wells. Impermeable liners and underdrain systems are typically required to prevent impacts. It is not recommended to have the reservoir course within 3 m (10 ft) of basements or building foundations unless adequate perimeter drainage, waterproofing, and geotechnical designs are provided and approved.

Efforts should be made to avoid having utility lines running through the reservoir course in no-infiltration system designs and designs where frequent fully saturated conditions are expected. If possible, utility lines should be located completely out of the pavement base or they should be adequately protected. In some cases, it may be practical to isolate utility lines situated in an open-graded base with flowable concrete fill. Designers should consult the local utility authorities to determine setbacks or casing requirements for site utilities. The implications of having greater amounts of infiltrating water and the potential effects on utility banks should be discussed with the utility suppliers, and approval should be requested prior to moving forward with final design.

Land Use and Contamination Potential

As stated in the NCDENR (2012) *BMP Manual*, "Permeable pavement should not be used to treat stormwater hotspots—areas where concentrations of pollutants such as oils, grease, heavy metals, and toxic chemicals are likely to be significantly higher than in typical stormwater runoff." Examples of stormwater hotspots included in the NCDENR manual include:

- Fueling facilities
- Commercial car washes
- Fleet storage
- Public works yards
- Trucking and distribution centers
- Highway maintenance facilities
- Vehicle maintenance areas
- SIC code heavy industries, such as airport maintenance areas

- Railroads and bulk shipping
- Race tracks
- Landfills/Solid waste facilities
- Wastewater treatment plants
- Scrap yards
- Brownfields and similarly regulated properties containing hazardous materials

Consistent with the EPA's regulations and as mentioned in the NCDENR (2012) *BMP Manual*, only the portions of the site where the industrial related activities or land use occurs are considered hotspots. The use of permeable pavement applications may be considered for other non-hotspot portions of the site.

1.6   Preliminary Data Review and Desktop Assessment

Prior to the initiation of the structural and hydrologic design of the permeable pavement system—after the preliminary project/site conditions and constraints have been defined, and after design options have been evaluated—it is recommended that the design engineer complete a preliminary desktop assessment of the key hydrologic conditions of the site. This assessment includes gathering information about soil conditions and the pertinent information relative to the expected use of the pavement (e.g., traffic loads expected). This preliminary hydrologic and structural information review is typically performed as a desktop assessment followed by focused field evaluations to further determine site conditions. Recommended steps are as described.

Desktop Assessment of Field Conditions and Pavement Requirements

A desktop assessment should be conducted prior to initiating field evaluations of soils conditions or selecting the pavement location and hydrological design methods. In general, this initial assessment includes a review and identification of the following:

- Underlying geology and soils maps including any karst areas
- NRCS hydrologic soil groups: A, B, C, D, and information related to depth, permeability, and structural properties of the soils
- History of fill soil or previous disturbances or compaction of soils at the site
- Topographical maps and drainage patterns
- Floodplain mapping
- Location of critical resources (streams, wetlands, waterbodies) on or near the project site
- Location of subsurface and surface structures on or near the project site
- Areas of industrial activities or stormwater hotspots (existing and/or proposed)
- Current and future land uses draining onto the site (drainage area delineation)
- Location of separate and/or combined sewer/stormwater systems and delineation of drainage areas, especially for green infrastructure retrofits
- Design storm with the return period and intensity in millimeters or inches per hour
  - Usually supplied by municipality or other regulatory agency
  - Rainfall intensity-duration-frequency or isohyetal maps also can be referenced to estimate the design storm depth
- Any proposed increase (percent) of run-on from adjacent impervious surfaces draining onto the permeable pavement. Direct discharges to the reservoir course, for example from rooftop downspouts can be included in a design, and must also include a detailed evaluation of the capacity of the entire system to accept additional runoff.
- The volume of runoff or peak flow to be captured, infiltrated, and/or released from the design storm
- An estimate of the vehicular traffic loads expressed as 80 kN (18,000 lb) equivalent single axle loads (ESALs) or Caltrans Traffic Index (TI) over the design life of the pavement, typically 20 years. Traffic load evaluations pertaining to the structural design of permeable pavements is discussed in the following section.

Field Analysis

After the preliminary desktop analysis for soil and site conditions is complete, a detailed field analysis program should be initiated. A soil sampling and testing program should be designed and supervised by a licensed professional civil or geotechnical engineer familiar with the local soils and pavement design. This engineer must determine soil design strength, permeability, compaction requirements, expected traffic type and loading, and other appropriate site assessment information.

A typical program for designing a traffic bearing permeable pavement system includes soil sampling and analysis, described as follows.

Soil Subgrade Sampling and Analysis Outline

1. Begin by marking locations of utility lines on the site. Utility companies and/or licensed professional utility marking services must verify locations.

2. For retrofit scenarios, determine the amount of asphalt or existing impermeable surface, if applicable, that will require removal.

3. Complete test pits (dug with a backhoe) to evaluate soils (recommend one pit for every 700 m² (7,000 sf) of paving with a minimum of two test pits per site). All pits should be dug at least 1.5 m (5 ft) deep with soil logs recorded to at least 1 m (3 ft) below the bottom of the base/subbase. More test pits at various soil depths (horizons) may be required by the engineer in areas where soil types may change, near rock outcrops, in low lying areas or where the water table is likely to be within 2.5 m (8 ft) of the surface.

4. Note any evidence of a high water table, impermeable soil layers, rock, or dissimilar layers that may require the design of a no infiltration system.

5. Classify soil types and complete soil tests to quantify soil permeability rates and structural properties. The following tests are recommended on soils from the test pit, especially if the soil has clay content. Other tests may be required by the design and/or geotechnical engineer. AASHTO tests equivalent to ASTM methods may also be used.

   - Identify the complete unified (USCS) soil classification using the test method in ASTM D2487.

   - Complete onsite soil permeability tests using local, state, or provincial recommendations for test methods and frequency. All tests for permeability (determining infiltration capacity) should be done at the elevation corresponding to the design location of the soil subgrade (at the bottom of the permeable base).

   - A recommended soil infiltration test is ASTM D3385, Test Method for Infiltration Rate of Soils in Field Using a Double-Ring Infiltrometer. Note that ASTM D5093 Test Method for Field Measurement of Infiltration Rate Using a Double-Ring Infiltrometer with a Sealed Inner Ring is appropriate for soils with an expected infiltration rate of 10^{-7} m/sec (1.4×10^{-2} in./hr) to 10^{-10} m/sec (1.4×10^{-5} in./hr). Percolation test results for the design of septic drain fields are not suitable for the design of permeable pavement systems.

6. In brownfield locations, highly urbanized areas, or other areas where contaminants may be present in the subsoil, it is likely that testing the soil for potential hazardous wastes or constituents is required, and is recommended to complete testing regardless.

Evaluation of Soil Sampling Results

It is critical to note that results from field permeability tests are approximations because the structure and porosity of soils are easily changed. On-site tests do not account for the loss of the soil's permeability from construction, compaction and clogging from sediment, nor do they account for lateral drainage of water from the soil into the sides of the base/subbase. Individual test results should not be considered absolute values directly representative of expected rates for the drawdown of water from the open-graded base/subbase. Instead, infiltration test results should be interpreted with soil texture, structure, and pore geometry. In-situ testing is recommended over laboratory soil infiltration tests because they are more representative of actual site conditions. Changes in the soil structure make it difficult to obtain representative sample for laboratory analysis.

For design purposes, a safety factor of 2 should be applied to the average or typical measured site soil infiltration rate. For example, a site infiltration rate of 25 mm/hr (1 in./hr) is halved to 13 mm/hr (4×10^{-6} m/sec or 0.5 in./hr) for design calculations. The safety factor helps compensate for decreases in infiltration during construction and over the life of the permeable pavement. A higher factor of safety may be appropriate for sites with highly variable infiltration rates due to different soils or soil horizons.

Soils with field-tested permeability equal to or greater than 4×10^{-6} m/sec (0.5 in./hr) usually will be gravel, sand, loamy sand, sandy loam, loam, and silt loam. These are soil types with usually no more than 15% passing the 0.075 mm (No. 200) sieve, which are typically characterized as NRCS A and B hydrologic group soils.

Silt and clay soils from the NRCS C and D groups will have lower permeability and will likely require partial- or no-infiltration designs using underdrains. Compacted clay soils will often have infiltration at 3.5×10^{-7} m/sec (0.05 in./hr) or lower. Even at this low rate, about 30 mm (1.2 in.) of water can infiltrate in 24 hours or more over longer time frames. For example, 25 mm (1 in.) of water can be infiltrated over 72 hours into a soil with a design infiltration rate of 0.33 mm/hr (9.2×10^{-8} m/sec or 0.013 in./hr). Therefore, clay soils—even when compacted—may still be capable of infiltrating some water. However, underdrains and overflow conveyance must be designed to handle the flows beyond the design capacity.

While soil subgrades with high infiltration rates typically have less than 5% passing the No. 0.075 mm (No. 200) sieve size, clay soils with up to 25% passing may infiltrate adequately depending on site conditions and specific characteristics. Soil permeability lower than 4×10^{-6} m/sec (0.5 in./hr) can be used to infiltrate water as long as the soil remains stable while saturated and when vehicle loads are applied. Soil stability under traffic should be carefully reviewed for each application by a qualified geotechnical or civil engineer. Pedestrian applications not subject to vehicular traffic can be built over soils with a lower permeability.

If soils have a 96-hour soaked California Bearing Ratio (CBR) <4% or are highly expansive, they will likely have very low permeability (Smith 2011). Designers must consider soil stabilization techniques to increase their stiffness under traffic loads. Soils treatments include mixing them with cement, lime, or lime/fly ash to control expansive soils and raise the CBR, but will have a corresponding reduction in permeability and thus infiltration capacity.

Results of Soil Compaction Results and Prevention

Pedestrian applications should not require soil subgrade compaction and it should be avoided if possible for vehicular applications. As a general rule, most installations will be over uncompacted soil subgrade. For vehicular applications, this subgrade layer should be evaluated by a qualified civil or geotechnical engineer for the need for compaction. If compaction is needed, on-site infiltration tests using previously cited ASTM standards should be conducted on test areas of compacted soil.

Diligent use and on-site control of tracked construction equipment traversing the soil subgrade will minimize inadvertent compaction. Wheeled construction equipment should be kept away from the exposed soil as it tends to concentrate loads, stress, and compaction.

1.7   Hydrologic and Structural Design

Permeable pavements serve a dual purpose as a stormwater management practice and as a load bearing surface. When designing a permeable pavement system, both the structural and hydrological requirements must be considered. The thickness of the permeable pavement and reservoir layer must be sized to support structural loads over saturated soils and to temporarily store the design storm volume (e.g., the water quality, channel protection, and/or flood control volumes). Each of the components in the permeable pavement cross-sections (**Figures 1-5 through 1-7**) must be specified for the type of pavement, use of pavement, region of use, site conditions, and goal(s) for using the permeable pavement. No single specification fits all purposes.

The flow chart or decision tree shown in **Figure 1-14** presents the structural and hydrologic design analysis process. Numerous variables, including soil types and pavement materials, will influence the design of the pavement structure.

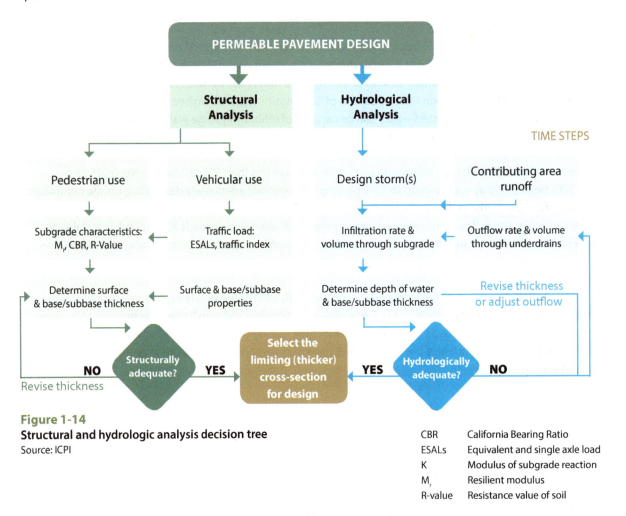

Figure 1-14
Structural and hydrologic analysis decision tree
Source: ICPI

CBR	California Bearing Ratio
ESALs	Equivalent and single axle load
K	Modulus of subgrade reaction
M_r	Resilient modulus
R-value	Resistance value of soil

Hydrological Design

Hydrological design generally relies on the following variables:

- Design storms and rainfall depth, typically issued by the local regulatory body/agency
- Run-on from surrounding areas
- Long-term soil infiltration rate, estimated from field measurements with an appropriate safety factor added by the designer
- Base/Subbase reservoir thickness and storage capacity
- System outflow configuration

The stormwater quantity entering the pavement surface can be characterized as a water balance among sources and destinations. Dynamic computational methods use small time steps to estimate the expected water inflow from precipitation and any surrounding areas that drain onto the permeable pavement. Infiltration, subsurface outflow, and possible surface runoff during each time step are also estimated.

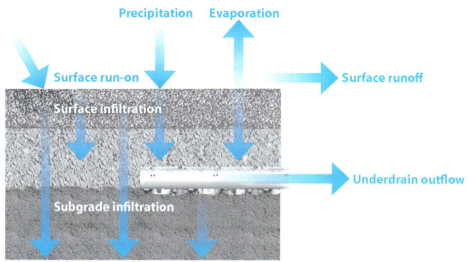

Figure 1-15
Water balance variables for permeable pavement
Source: © VHB

Several standard computational methods are available, and these or other locally developed methods may be required by state or local regulatory agencies. A summary of the more commonly used hydrologic methods pertaining to permeable pavement design are discussed in Chapter 9. Figure 1-15 illustrates the critical variables that should be modeled in any computational method for permeable pavement, regardless of the specific computational method used.

Run-on From Surrounding Areas

Water entering the permeable pavement surface and from surrounding areas is estimated using rainfall, soil type, land use, and runoff characterizations. A commonly used rainfall and runoff characterization method is the NRCS Technical Release 55 Urban Hydrology for Small Watersheds. NRCS design storms tends to overstate the rainfall (characterized as synthetic events), which may lead to conservative hydrologic designs. While this is a commonly accepted method, using local rainfall data is a better approach.

All inflow from adjacent areas is assumed to be sheet flow onto permeable pavement. The permeable pavement surface has a finite inlet capacity. Pavement surface infiltration rates can be input by the user based on test results or experience.

Rainfall/Design Storm

Rainfall events can be selected based on local design requirements. Rainfall timing can be important when evaluating permeable pavement potential to infiltrate water from surrounding areas. The time delay between the rainfall on these areas (with some infiltration) and the time the water enters the permeable pavement surface during the peak rainfall intensity can also reduce the peak outflow, thereby conserving the need for larger storm drain system pipes and reducing potential downstream flooding or erosion.

Water Outflow

Stormwater infiltrates from the permeable pavement base/subbase into the soil subgrade, groundwater, and/or flows to an underdrain system. Evaporation and transpiration that occur with paver/vegetation systems are factors to include in the modeling of the system particularly in arid climates, especially during the hot season. The designer needs to carefully estimate the amount of water infiltrating the soil subgrade and into underdrains. Underdrains are typically used over low-infiltration soils, and the designer specifies the pipe size, slope, horizontal spacing, and drainage height above the soil subgrade. The height of the underdrains can be designed to specifically create some water detention in the reservoir course to encourage increased infiltration. This process may also promote nutrient reduction through de-nitrification.

Soil Subgrade Infiltration

Measured or referenced soil subgrade permeability should be a saturated conductivity that yields a calculated infiltration rate over time. The simplest approach uses Darcy's Law (Cedegren 1989). The Green-Ampt equation can be used as well. Since the water table is typically some distance below (minimum separation recommendation typically 0.61 m [2 ft]) the base/subbase layer, the hydraulic gradient can be assumed to be 1.0 as the drop in elevation causes downward flow. Most designs assume that water infiltration into the soil subgrade occurs uniformly across the bottom of the pavement as the base/subbase becomes saturated. Additionally, lateral flows through the open-graded aggregates are assumed to be a portion of the rate with no aggregate. The designer exercises some latitude in deciding lateral flow rates within base/subbase aggregates.

Since predicting sediment loading on the soil subgrade and reduction of infiltration is difficult, the previously noted conservative infiltration reduction factor of 0.5 (safety factor of 2) can be applied to account for potential clogging and reduction of the soil subgrade infiltration rate. As the water depth in the base/subbase increases, the static pressure also increases, affecting the drainage rate.

Permeable pavements may be designed to detain water, which can assist in nutrient reduction. This approach is more amenable in low-infiltration rate soils, which can also capture metals. In such cases, detention pond design principles can be applied to inflow, storage, and outflow calculations. The maximum resident time for water storage generally should not exceed 72 hours, including the storm duration, but must follow local regulatory requirements, typically determined by regional rainfall patterns. Excess water that cannot be contained by the base/subbase could exit to swales and/or to down-gradient bioretention areas, catch basins, and/or storm drains. Besides detention that encourages denitrification, additional nutrient treatment can be realized with release to other stormwater management Best Management Practices (BMPs).

The USDA's NRCS (formerly Soil Conservation Service) curve numbers (CN) are often used to characterize runoff in drainage system design. The CN is a dimensionless number that characterizes the relationship between runoff, land use, and soil type. CN values range between 0 and 100. The 100 value means a completely impervious surface and zero (0) value means a surface capable of taking in all water even in flood conditions, which is practically impossible.

A core concept in calculating a CN is the amount of water stored and infiltrated by a parcel of land from a rain event that does not eventually become runoff, also known as initial abstraction. CNs will vary among permeable pavement designs and with the soil types under them. CNs for permeable pavements can range in the mid-40s on sandy soils to the low-80s on practically impervious soils. The designer establishes the total amount of water infiltrated to calculate a CN. Further information on the strengths and weaknesses of using CNs are found in the literature by Hawkins et al. (2009) and Ballestero (2011).

Permeable pavement can also be designed to address, in whole or in part, the detention storage needed to comply with channel protection and/or flood control requirements. The designer can model various approaches by factoring in storage within the stone aggregate layer, expected infiltration, and any outlet structures used as part of the design. Routing calculations can also be used to provide a more accurate solution of the peak discharge and required storage volume. Some of the key hydrologic factors and benefits from permeable pavements are noted as follows.

Management of Small Storms

Permeable pavement typically manages small storms extremely well and provides water quality treatment, peak flow, and volume reductions. The reduction of volumes and peak flows from frequent small storms contributes to channel protection as well as the attainment of total maximum daily load (TMDL) goals.

Control of Larger Storms

Larger reservoir volumes and high permeability soils can result in significant control of volumes and peaks from larger storms depending on the design. Overflow design features must be included for every system when extreme intensities and durations exceed the capacity of the system as designed.

Peak Flows

Peak flow reduction can be achieved for small storms and larger storms, and the replacement of existing pavement with permeable options will show significant reductions in peaks. If underdrains or other controlled discharge elements are used, the hydrologic analysis should include the results of the discharge typically in the same fashion as modeling a detention basin.

Storage Within Pavement or Underlying Layers

Storage is typically calculated by identifying the void space in the reservoir area and applying this as a static volume or as a larger volume accounting for the dynamic process of infiltration simultaneously occurring during rainfall accumulation. Computation models can assist in conducting sensitivity analyses that balance these factors in the design.

Storage Capacity

The total storage capacity of the permeable pavement system should not include the capacity within the surfacing and the bedding layer, if present, in the case of PICP. Storage is credited to the void space in the base and subbase layers, and capacity may be increased with optional underground chambers. The computation of total volume of runoff captured should include the amount of water that leaves the system through infiltration during the storm into the underlying soil.

Another potential source of storage is available with curbed pavement systems. This design is only recommended in warm climates where freeze/thaw does not occur. A design incorporating captured water up to the depth of the curbs is not normally included in commercial parking lots, on-street parking areas or pedestrian areas from an intense storm. This design feature may be included in permeable pavement. However, this approach may not have much utility in applications such as parking areas with well draining soils with brief water impoundment prior to complete infiltration. For slower draining soils, a curb adds to the storage capacity of the pavement system.

Structural Design

For structural design of conventional, impervious roads, and base/subbase, many local, state, and federal agencies use design methods published by AASHTO (1993) *Guide for Design of Pavement Structures*. While the AASHTO methodology is familiar to many civil engineers, stormwater agency personnel who are not actively involved with pavement design are encouraged to familiarize themselves with AASHTO as it is referenced in permeable pavement specifications and design recommendations for local, state, or federal BMP/LID manuals and regulatory documents.

State and federal highway engineers are moving toward the AASHTO (2004) *Mechanistic-Empirical Pavement Design Guide (ME PDG) for New and Rehabilitated Pavement Structures*, which relies on mechanistic design and modeling, i.e., analysis of loads, resultant stresses, and strains on materials and the soil subgrade. The ME PDG model was developed and calibrated by state and federal highway agencies across a wide range of highway loads, load testing, soil types, and climatic conditions. This model has not been calibrated for permeable pavements subject to significantly less traffic loads and constructed with open-graded, crushed stone bases/subbases.

Many local transportation agencies use the AASHTO (1993) Guide. The concepts in it emerged from accelerated vehicle load testing of flexible asphalt and rigid concrete pavements in the 1950s. Repeatedly loaded with trucks, the test results established relationships between loads and pavement damage, expressed as rutting in flexible pavements and cracking in rigid pavements, as well as characterizing the strength of surface and base/subbase materials. The result was empirical AASHTO equations for flexible and rigid pavement design.

For flexible pavement, the AASHTO (1993) equation in the Guide calculates a structural number (SN) given traffic loads, soil type, climatic and soil moisture conditions. The designer finds the appropriate combination of pavement surfacing and base/subbase materials whose strengths are characterized with layer coefficients. When added together, the coefficients (representing various pavement material thicknesses) should meet or exceed the SN required for a design. This empirical design approach appears to be applicable to permeable pavement; however, some modification to layer coefficients may be necessary.

The structural design process will vary according to the type of pavement selected. Flexible pavement design concepts in AASHTO (1993) can be applied to porous asphalt and permeable interlocking concrete pavements. The concrete industry has adapted their "StratPave" rigid pavement design procedure for the structural design of pervious concrete. Like the permeable interlocking concrete pavement industry, the pervious concrete industry provides (empirical) base/subbase thickness design charts that rely on soil strength, base/subbase, and traffic load inputs. Both industry associations offer software programs for structural as well as hydrologic design. Designers should refer to individual chapters for specific information on structural design for each pavement type.

For vehicular applications, designers must consider structural loading requirements for pavement based on the anticipated loads. The pavement surface must be able to support the maximum anticipated traffic load of the pavement. Once a base/subbase thickness is determined for structural support, the hydrological and structural base/subbase thickness is determined by selecting the resulting thicker base.

Hydrological design is a straightforward task for stormwater managers and remains consistent regardless of the permeable surfacing selected. However, structural (base thickness) design varies with each system. Designs are restricted to moderate vehicular loads due to a small body of knowledge and experience on structural performance of open-graded aggregate bases/subbases and larger heavier loads.

The structural design of permeable pavements involves consideration of four main site elements.

1. **Total traffic**—Determine the anticipated traffic load/type for the particular pavement applications.

    - Determine if the pavement be used for vehicular or pedestrian traffic

    - Estimate expected vehicle weight

    - Estimate the expected traffic loads (average daily trips)

    - Define the design life of the pavements

    - Calculate the total equivalent single axle load (ESALs) or Caltrans TI (for porous asphalt and PICP) or axle load spectra (for pervious concrete)

2. **Soil properties**—Determine soil type and physical properties.

    - Determine the load bearing capacity—Estimate the bearing capacity of the underlying soils using 96-hour soaked California Bearing Ratio (CBR), modulus of subgrade reaction (K), R-value test, or resilient modulus (M_r)

    - Determine soil compaction—Determine if the soils need to be compacted. Designers should note that if the underlying soils have a low CBR (<4%), the soils may need to be compacted to at least 95% of standard Proctor density

3. **Environmental factors**—Estimate how frequently the pavement subgrade may become saturated and estimate how quickly is it expected to drain

    - Adjust the soil strength or bearing capacity characterization based on the amount and frequency of rainfall and the ability of the underlying soils to infiltrate water

4. **Surface, bedding, and reservoir layer strength coefficients and design depth**—Determine the required depth of the pavement layers needed to support traffic loads

    - For porous asphalt and permeable interlocking concrete pavement, determine the pavement structural number (SN) based on:

        - The underlying soil strength and the anticipated traffic loads

        - The layer coefficients of the pavement layers so that the SN is met or exceeded

    - For pervious concrete, apply rigid pavement design methods with allowances for reduced pervious concrete flexural strength

1.8   Water Quality

Current studies indicate favorable-to-excellent water quality benefits from permeable pavement. The pavement surface's role in reducing pollutants from the base and subbase has been well documented. Recent findings for pollutant removal capabilities for various permeable pavements are included in **Appendix B**.

As noted earlier, the addition of a filtering course within the pavement subbase with specifically selected filtering media can significantly increase pollutant load reductions for individual contaminants of concern as well as other contaminants included in the runoff. However, the added expense may justify the use of other water quality treatment tools such as bioretention basins or water quality treatment swales. Recently documented highlighted water quality benefits include:

- **Pollutant concentrations**—Permeable pavement decreases concentrations of heavy metals, motor oil, sediment, including TSS, and some nutrients (Bean 2007; Brattebo 2003; Fassman and Blackbourn 2011; James 1994 and 1996; Pratt 1995 and 1999).

- **Pollutant loads**—Permeable pavements lower total pollution loadings over standard pavements because the overall total volume of runoff is much lower (Bean 2007; Day 1981; Fassman and Blackbourn 2011; Rushton 2001). Most heavy metals are captured in the top layers of permeable pavements (Colandini 1995; Dierkes 2002).

- **Thermal**—Permeable pavements can cause a reduction of thermal pollution (James 1996; Karasawa 2006) compared to conventional asphalt.

- **Buffering**—Permeable pavements can buffer acidic rainfall pH (Collins 2008; Dierkes 2002; James 1996; Pratt 1995), which is likely due to the presence of calcium carbonate and magnesium carbonate in the concrete pavement and aggregate materials. Pervious concrete provides a greater buffering capacity than conventional asphalt due to the cement and contours in the pavement geometry and the additional coarse aggregate layer through which water migrates.

- **Phosphorus**—Some permeable pavement studies have shown removal of total phosphorus (P) (Bean 2007; Clausen 2006; Day 1981), often attributed to adsorption to sand and gravel base/subbase materials. Similar studies have observed little change in TP concentrations of permeable pavement drainage (Bean 2005; Collins 2008; James 1996). The addition of engineered aggregates (i.e., coated with nutrient-reducing substances) or use of iron in sand filters may provide significant phosphorus control over non-engineered aggregated. Engineered aggregates technology is discussed in Chapter 6.

- **Nitrogen**—Several studies suggest that aerobic conditions, which result as permeable pavement drains, can result in nitrification of ammonia-nitrogen (NH_4-N) to nitrate-nitrogen (NO_3-N). Compared to asphalt, substantially lower NH_4-N and total Kjeldhal nitrogen (TKN) concentrations and higher NO_3-N concentrations in permeable pavement drainage have been measured in multiple experiments (Bean 2007; Collins 2008; James 1996). A few studies show decreased concentrations of all measured nitrogen species (NH_4-N, TKN and NO_3-N) (Clausen 2006; Pagotto 2000). The creation of an anaerobic zone in the pavement reservoir may encourage denitrification and enhance total N removal similar to the well-established process in detention ponds.

- **Total suspended solids (TSS) particulate load**—Particulate contaminants in stormwater do not travel through the pavement and contaminate groundwater. Long-term studies and simulations of permeable pavement pollutant distributions have revealed low risks of subsoil pollutant accumulation and groundwater contamination (Dierkes 2002; Legret 1999; Van Seters 2007).

- **Salts**—Soluble contaminants such as deicing salts are not treated, but there is typically a reduced need for their application as the permeable surface takes in the melted ice and snow (Houle 2009).

- **Spills**—Hazardous spills on permeable pavement result in the same situation as spills into a detention or infiltration basin if used with standard form conventional impervious pavements. Therefore, there is no increased risk for such spills into permeable pavements. Such spills can be completely contained and lend themselves to direct remediation when compared to the greater complexities involved in remediating hazardous pollutants passed from impervious surfaces into a stream or lake or municipal stormwater system.

- **Litter/Debris/Nutrients as surface accumulations**—Most permeable pavements must be vacuumed periodically, and this results in removal of many other contaminants such as sediment as well as litter that can diminish water quality. Maintenance and surface cleaning are covered in Chapter 8.

1.9   Sustainable Design Credits

Permeable pavements can contribute towards green building certification. The following benefits are highlighted within the US Green Building Council's (USGBCs) Leadership in Energy and Environmental

Design (LEED) program. **Table 1-2** outlines LEED credits that apply to the major permeable pavement types covered in this report. Other rating systems provide credits for permeable pavement use, including SITES Version 2. Permeable pavements that use life cycle assessment to quantify energy use and other environmental impacts can qualify for credits in sustainable road rating systems. These rating systems include the University of Washington's GreenRoads, the Federal Highway Administration's (FHWA) sustainable highways evaluation tool called INVEST, the Ontario Canada Ministry of Transportations Greenpave Program and the Institute for Sustainable Infrastructure's Envision program.

Table 1-2 US Green Building Council Credits in the LEED® v4 Building Design and Construction Rating System

CREDIT CATEGORY	POINTS			
	Porous Asphalt	Pervious Concrete	PICP	Grids
Sustainable sites				
Construction activity pollution prevention	Prerequisite (no points)			
Environmental site assessment	Prerequisite (no points)			
Open space	1	1	1	1
Rainwater management*				
95th percentile storm	2	2	2	2
98th percentile storm	3	3	3	3
Zero Lot Line—85th percentile storm	3	3	3	3
Heat island reduction: Non-roof measures*		2	2	2***
Water efficiency				
Outdoor water use reduction	Prerequisite (no points)			
Indoor water use reduction	Prerequisite (no points)			
Outdoor water use reduction*				
No irrigation required	2	2	2	2
Reduced irrigation	2	2	2	2
Indoor water use reduction**	1 to 7	1 to 7	1 to 7	1 to 7
Materials and Resources				
Building product disclosure and optimization				
Environmental product declarations	1	1	1	1
Multi-attribute optimization^A	1	1	1	1
Raw materials source and extraction reporting	1	1	1	1
Leadership extraction practices^A	1	1	1	1
Material ingredient reporting	1	1	1	1
Material ingredient optimization^A	1	1	1	1
Product manufacturer supply chain optimization^A	1	1	1	1

Note: See the LEED v4 Neighborhood Development rating system for credits specific to neighborhood design. LEED v4 rating systems offer additional points through Regional Priority and Innovative Design credit categories.
*Healthcare facilities are reduced by one point.
**Schools, retail, hospitality and healthcare facilities earn 1 to 5 points.
***Grids with topsoil and grass surfaces; all other grid applications qualifying for Rainwater Management and Water Efficiency credits must use open-graded aggregates within the grid surface and in the base/subbase.
^AThese credits include a two-fold location valuation factor to incentivize purchases that support the local economy. To qualify, materials and products must be extracted, manufactured and purchased within 100 miles (160 km) of the project.

Source: ICPI

1.10   Implementation

All pavements require proper design, construction specifications, installation, and maintenance. Additional construction requirements for permeable pavements include: (1) protection of aggregate and the finished surface from sources of clogging, (2) regular vacuum cleaning after construction and before site stabilization, and (3) possibly vacuuming to restore clogged surfaces that do not receive periodic vacuuming over many years. General recommendations for installation are provided in this chapter with more detailed information for each pavement in the Chapters 2 through 6 to follow. Chapter 8 is dedicated to maintenance requirements applicable to all pavements. The key factors critical for proper performance of permeable pavements are summarized as follows.

Testing for Soil Permeability

As outlined in the previous section, permeability tests should have been performed on the soil subgrades prior to the permeable pavement design. The tests should have been completed at the elevation for which natural soil subgrade infiltration is being proposed. Permeable pavements can be used over low permeability soils, but design features such as underdrains or a raised drain/outlet structure with greater reservoir storage need to be included for proper system functionality. While this step should have been completely vetted prior to reaching the installation phase, the information should be reviewed and confirmed prior to installation.

Pre-Construction Meeting

Project specifications should include the requirement for a pre-construction meeting. This meeting should be held to discuss methods of accomplishing all phases of construction operations, contingency planning, and standards of workmanship. For permeable pavement, specific focus on protecting the pavement and base/subbases during all stages of construction is critical. The general contractor typically provides the meeting facility, meeting date and time.

Attendees: It is recommended that representatives from the following entities attend:

- Contractor superintendent
- Permeable pavement subcontractor foreman
- Paving material manufacturer's representative
- Testing laboratory(ies) representative(s)
- Engineer or representative

Agenda: It is recommended that the following items be discussed and determined:

1. **Soil permeability results**
2. **Set backs** (building foundations, water supply, etc.)
3. **Sediment control/pavement protection**—Methods for keeping all materials free from sediment during storage, placement, and on completed areas
4. **As-built controls**—Methods for checking slopes, surface tolerances, and elevations
5. **Materials management**—Paving materials delivery method(s), timing, storage location(s) on the site, staging, paving start point(s), and direction(s)

6. **Paving schedule**—Anticipated daily paving production and actual record

7. **Design details**—Joints, edging, infrastructure, overflow, monitoring wells, cleanouts, etc. Review details of pavement installation related to joining with adjacent pavements, land uses, etc.

8. **Testing**
   - Testing intervals for sieve analyses of aggregates and for pavement materials
   - Field testing method(s) for identifying pavement material batches or packages delivered to the site and after placement
   - Testing lab location, test methods, report delivery, contents, and timing

9. **Regulatory requirements**—Stormwater permits (MS4, Construction General Permit); natural resources (wetlands and forests); soil characterization and management (as applicable)

10. **Inspections**
    - Inspection checklists (see **Chapter 8**, **Checklist 3: Annual Permeable Pavements Inspection**)
    - Engineer inspection intervals
    - Procedures for correcting work that does not conform to the project specifications

11. **Test panel (mock-up) location and dimensions**—A test panel should be made if required per the specifications and/or if appropriate for site size/application (refer to specific pavement type chapter, manufacturer, and design team for input regarding the need for a mock up).

Proper Installation

Different types of permeable pavements require different installation techniques, but all require protection of the pavement and base/subbases during installation, testing of materials to ensure they match the design specifications and monitoring in the field to ensure compliance with depths of materials and details are adhered to (**Figure 1-16**).

Figure 1-16
Installation of porous asphalt parking lot, North Shore MA
Source: VHB

Manufactured pavers, pervious concrete, and some proprietary pavements are often installed by trained and/or certified installers. Some provide a warranty. For porous asphalt and pervious concrete, the specifications for materials and installation requirements (climate concerns etc.) may differ from region to region. It is recommended that trained or certified installation crews that specialize in the pavement type are used to ensure industry-accepted best practices are applied. Proper installation, of course, begins with technically correct details and specifications, followed by proper testing of materials and field inspections by the design engineer.

Observation Well

Large-scale permeable pavement installations should include one or more monitoring/observations wells and cleanouts. Wells are used to observe the rate of drawdown following storm events. Typical design is a 100 to 150 mm (4 to 6 in.) diameter perforated pipe that extends vertically to the bottom of the reservoir layer (Figure 1-17). A lockable cap or lock box, can be flush with the surface or beneath pavers if not asphalt or concrete. The depth to the soil subgrade should be marked on the lid. Locate the well in the furthest down-slope area with a minimum of 1 m (3 ft) distance from the pavement edge. If final surface paving is not to be completed within a short time frame, additional stabilization and security of the well should be completed once placed in the subgrade. If final surface paving is not completed within a short time frame, it will be necessary to protect the well from construction equipment damage.

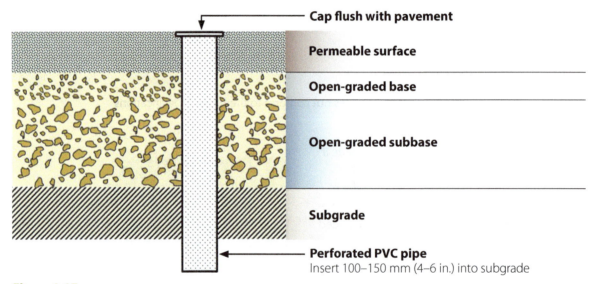

Cap flush with pavement

Permeable surface

Open-graded base

Open-graded subbase

Subgrade

Perforated PVC pipe
Insert 100–150 mm (4–6 in.) into subgrade

Figure 1-17
Surface inspection monitoring well with covered cap
Source: Applied Research Associates

The level of water in the monitoring well should be observed during construction and regularly over the service life of the pavement to ensure that the pavement base/subbase is draining as designed by the engineer and in compliance with applicable local requirements. The draindown rate of water should be observed at least twice per year after a major storm event.

Construction Inspection and Testing

On-site quality control is essential for materials delivered, materials placement, and materials after placed. Industry guidelines and tests should be included in the specifications and followed. Tests performed on porous asphalt or pervious concrete materials during construction are necessary to ensure compliance with plans and specifications. Some standard tests and references for construction installation checklists and testing requirements are included in Chapters 1, 3, 4, 5, 8, and Appendix C as well as the appendices of the Rhode Island (2010) *Stormwater Design and Installation Standards Manual*. Note that materials and practices vary from region to region.

Pavement Surface Protection

Erosion control is critical to avoid surface or subgrade clogging. Recommended methods include using staged construction, temporary roads, restricting trucks with muddy tires or deploying a tire washing station.

The individual chapters for specific pavement types that follow this section include more detailed recommendations for the design and construction implementation of each pavement type. General recommendations for implementing the construction phases of permeable pavement projects are provided in the **Checklist 2: Recommendations for Permeable Pavement Construction Procedures Summary** at the end of this chapter.

1.11   Maintenance

Studies show that permeable pavement surfaces will continue infiltrating if their surfaces are vacuumed regularly. Sweeping is typically completed with regenerative vacuum sweepers at a recommended minimum of twice per year. If the surfaces become clogged due to lack of regular surface cleaning, they can recoup some infiltration through use of a regenerative air vacuum sweeper as defined in **Chapter 8** (Chopra 2007; James 1996; Vancura 2010). Vacuum maintenance can prevent full surface replacement. Prior to recommending permeable pavement, the designer should consider whether the location can be protected during construction and post construction from clogging sources. In addition, the project owner should be advised of the biannual surface cleaning requirements.

Pavement Cleaning

While cleaning twice per year is recommended for typical applications, this schedule can be adjusted as needed. Regenerative air vacuum machines are recommended for regular cleaning, and pure vacuum type units are recommended only for restoration of severely clogged surfaces that have not been vacuumed regularly. Simple broom sweepers are not recommended since they do not collect and remove accumulated debris. All inlet structures within or draining to the reservoir under the permeable pavement should also be cleaned on a biannual basis or as needed. An inspection well should be included in the design in order to monitor whether the base/subbase is draining as designed. Detailed information on maintenance, including cold climate considerations, is included in **Chapter 8**.

Sediment, Run-on and Organic Matter Control

All permeable pavements are susceptible to clogging from sediment, fines, and organic matter that may be deposited on the pavement. A practical but rigorous erosion and sediment control program should be developed to protect the pavement. The program should include a minimization of potential sources of sediment that can be deposited/conveyed to the system (such as disturbed up-slope areas or tracked on via vehicles). Adjacent vegetated areas must be well maintained (and, ideally, sloped away from the pavement) to prevent soil, mulch, etc., from washing or being deposited onto the pavement. If any material is deposited on the pavement, the material should be removed immediately to prevent clogging. If any bare spots or eroded areas are observed within adjacent planted areas, they should be immediately replanted and stabilized. Planted areas should be inspected at least twice per year to be sure they are not eroding onto the pavement. Practical recommendations for assessment and maintenance of permeable pavements can be found at http://stormwaterbook.safl.umn.edu/content/maintenance-infiltration. Details are also included in **Chapter 8**.

Signage

One or more signs should be placed in a visible location identifying the pavement as permeable and indicating prohibitions (**Figure 1-18**). The various prohibitions will be specific to the pavement design and use. This information should also be provided to all service contractors for the property that may engage in any activities including landscaping, snow plowing, or facilities-related activities that in effect use maintenance or storage on or near the pavement. Educational signage may also be provided in addition to signs regarding maintenance in use of the pavement.

Figure 1-18
Recommended signage for maintenance staff
Courtesy NCDENR

1.12   Costs

Permeable pavement system costs vary depending on the type of pavement surface as well as the base/subbase depth and design. Costs also vary depending on whether the practice is a retrofit or new construction. Prices for plant-prepared materials vary based on the distance from the supplier to the project site and whether the materials in the specifications are available locally. The material costs for permeable pavement systems are generally higher than standard asphalt. While permeable pavements may need additional depth of base or subbase materials to support anticipated vehicular loads compared to conventional pavements, this subbase typically includes a reservoir course for stormwater storage. The detained and/or infiltrated stormwater can result in site development and management cost reductions if stormwater detention or infiltration basins are not needed on the property. Cost savings related to the potential elimination of drainage infrastructure such as curbs, catch basins, and pipes may also be realized when utilizing permeable pavements instead of typical impervious pavement installations. These systems typically retain vehicle or pedestrian function on a site. The cost benefit of these systems should consider the combined functionality and stormwater management provided by these systems. Estimated costs for permeable pavements are shown in the **Table B-4** in **Appendix B**.

1.13   Checklists 1 and 2

The permeable pavement designer must first assess the appropriateness of permeable pavement for the project site. If such pavement satisfies the site design requirements, the next step is evaluating specific surfacing types, related hydrologic, and structural designs, details, specifications, installation, and maintenance over the pavement's service life. The **Checklist 1: Design Considerations Common To All Permeable Pavements Summary** evaluates some key design elements, and **Checklist 2: Recommendations for Permeable Pavement Construction Procedures Summary** lists construction considerations. Both summary checklists are provided on the following pages.

CHECKLIST 1 ☑

Design Considerations Common To All Permeable Pavements

☐ 1. REGULATION (Check Requirements and Guidelines)

☐ **a.** Determine if the local regulatory agency allow permeable pavements. If not, determine who can authorize approval.

☐ **b.** Determine if these pavements are prohibited in certain areas, such as groundwater recharge zones, karst geology and fill situations.

☐ **c.** Determine if there are credits offered to reduce stormwater utility fees, permitting fees, or reduced site development costs for using permeable pavements.

☐ **d.** Determine if there are there regulatory hydrologic control or water quality requirements associated with the use of permeable pavements.

☐ **e.** Determine if there are there water quality control requirements specific to permeable pavement use.

☐ **f.** Determine if there are specific design guidelines or specifications mandated under applicable federal, state, or local regulations.

☐ 2. SITE (Identify Site Conditions)

☐ **a.** **Groundwater elevation**—The bottom of the permeable pavement base/subbase should be at least 60 cm (2 ft) above the seasonal high groundwater level within the soil subgrade.

☐ **b.** **Groundwater supply**—Identify nearby groundwater supply wells or recharge zones and requirements.

☐ **c.** **Floodplains**—The use of permeable pavement in floodplain areas is generally not recommended due to an increased risk of clogging if sediments are deposited on the pavement during flood events.

☐ **d.** **Bedrock**—Identify bedrock elevations and/or karst geology. Bedrock directly under the permeable pavement base/subbase typically requires the use of an impermeable liner.

☐ **e.** **Soil properties**—Determine soil type and physical properties:

  - Soil Classification—Determine classification of soil borings or test pits on the site.
  - Soils Present—Identify soil types and estimate elevation of aquatard or low-permeability soils if present.
  - Load Bearing Capacity—Estimate the bearing capacity of the underlying soils (CBR, R-value or resilient modulus) and determine the soil support value. Determine requirements for intended vehicular traffic use.
  - Soil Compaction—Specify soil compaction requirements. If the underlying soils have a low California Bearing Ratio (CBR) (<4% soaked CBR), they may need to be compacted to at least 95% of standard Proctor density, which reduces their infiltration rate.
  - Soil Permeability—Identify soil permeability (hydraulic conductivity rate, K) and rate to be used for design and check with local requirements/regulations on methodologies and guidelines. For larger projects with adequate budgets. Identify low permeability soils and constraints.
  - Soil/Groundwater Contamination—Research/identify the presence of any soil or groundwater contamination and how it may affect design. Permeable pavements should not be used in areas of groundwater/soils contamination without an underdrain above the liner.

☐ **f.** **Rainfall**—Evaluate regional rainfall and estimate how frequently the pavement will be inundated and how quickly the pavement will drain based on the ability of the underlying soils to infiltrate water.

Source: © VHB

CHECKLIST 1 ☑

Design Considerations Common To All Permeable Pavements *(continued)*

☐ 3. DESIGN (Identify Design Conditions, Constraints and Needs)

☐ **a. Surface, bedding, and reservoir layer strength coefficients and design depth**—Determine the required depth of the pavement and layer thicknesses needed to support traffic loads.

☐ **b. Slope**—Identify existing slopes and design slopes. (Note: Recommended surface slope to be at least 1% and no greater than 5% unless special sloped design methodologies are applied for subgrade slope.)

☐ **c. Drainage area**—Identify if the pavement is taking any additional flows other than the rainfall on the pavement itself and if it is from rooftop or adjacent impervious areas. Consider whether slope of subgrade and infiltration rate would result in significant lateral flow and require the need for alternative designs.

☐ **d. Utilities**—Identify known underground utilities or easements for past/future utilities.

☐ **e. Setback from buildings**—Identify the presence of adjacent buildings, basements, foundations. Depending on soils, drainage area, pavement systems should be located a minimum of 3 m (10 ft) downslope from buildings and foundations. Permeable pavements adjacent to basements/ foundations should be designed to prevent impact from stormwater.

☐ **f. Setback from private wells**—Identify local requirements. Recommended minimum 30 m (100 ft) is common, but must be researched by design engineer for potential impacts.

☐ **g. Setback from septic systems**—Recommended minimum 30 m (100 ft) is common, but must be researched by design engineer for potential impacts.

☐ **h. Edge restraints**—Most pavement systems require edge restraints to render stationary edges. Include details and specifications for restraints.

☐ **i. Monitoring wells**—A capped well should be installed at the lower end of the facility for monitoring the length of time required to fully drain the reservoir course between storms. This is typically a 100 to 150 mm (4 to 6 in.) diameter perforated pipe with a screw cap.

☐ **j. Cleanouts**—Piped surface access to underdrains will facilitate removal of sediment that may accumulate.

☐ 4. CONSTRUCTION (Specify Construction Requirements*)

☐ **a. Prepare construction specifications**—Specific to region, site, traffic loads, and hydrologic requirements

☐ **b. Weather**—Permeable pavement should not be installed during periods of extreme weather. Identify weather-related limitations on installation conditions in the construction specifications. Note: Porous asphalt and pervious concrete will not properly pour and set in extremely high and near-freezing air temperatures.

☐ **c. Installation**—Require trained, experienced materials producers and construction contractors.

☐ **d. Sediment control**—Provide clear sediment control practices on plans and in construction specifications before and during construction, as well as plans for stabilization of bare soil during the time required to establish vegetative cover. In addition, provide maintenance specifications that address sediment removal from accidental erosion or spills. In most cases, permeable pavements are built at the end of a construction project and will require protection from other construction equipment, dirt tracked from tires and soil eroding onto the pavement. Whenever

CHECKLIST 1 ☑

Design Considerations Common To All Permeable Pavements *(continued)*

4. CONSTRUCTION (Specify Construction Requirements*) *(continued)*

possible, restrict construction traffic from permeable pavements, provide a temporary road or provide a protective cover of geotextile and aggregate that can be removed when construction is completely finished.

☐ **e. Inspections and testing**—Clearly define materials and site inspection requirements including testing methods in the specifications.

☐ **f. Monitoring wells**—A capped well should be installed to the lowest elevation of the facility, no closer than 1 m (3 ft) from the perimeter, for monitoring the length of time required to fully drain between storms. The well should be protected during construction.

Note: Construction specifications should define the minimum requirements outlined above. A detailed summary checklist, **Checklist 2: Recommendations for Permeable Pavement Construction Procedures Summary**, is provided following this checklist.

5. OPERATION AND MAINTENANCE (Long-term O&M)

(See **Chapter 8** for full details on maintenance requirements and **Checklist 4: Annual Permeable Pavements Maintenance**.

☐ **a. Inspections**—Annual inspections in the spring to provide continual infiltration/draining.

☐ **b. Surface cleaning**—Minimum twice per year, depending on use with commercial vacuum sweeper. Permeable pavements should not be washed with high-pressure water systems or compressed air units unless part of a specific rehabilitation process.

☐ **c. Heavy vehicles**—Trucks and other heavy vehicles should be prevented from tracking or spilling dirt/debris/materials on the pavement.

☐ **d. Hazardous materials transport vehicles**—These vehicles should be excluded from driving on pavement.

☐ **e. Drainage areas**—No vegetated, exposed soil areas, or unswept pavement areas should be included in the contributing drainage area and allowed to flow onto the pavement.

☐ **f. Grid paver systems planted with grass**—Should be mowed regularly with clippings removed. May require some watering or fertilization to establish growth (covered in detail in **Chapter 5**).

☐ **g. Winter maintenance**—No sand application. Moderate or no deicer use. Studies show permeable pavement requires as much as 75% less salt than conventional per season. Do not store plowed snow piles on the permeable pavement (UNHSC 2010; Roseen et al. 2013).

*This information is a suggested framework for a checklist and should not be considered an exhaustive or complete list of all items that should be reviewed. Advice and direction from a competent professional in the field should be sought for site specific application of any and all material included in this report.

Source: © VHB

CHECKLIST 2 ☑

Recommendations for Permeable Pavement Construction Procedures

☐ **1. COMPLETED SOILS TESTING**

☐ a. Verify that soils tests indicated in the specifications have been completed. Note any changes to proposed use, materials or design that may have been made or need to be made as a result of the specified soils tests.

☐ **2. CONDUCTED PRE-CONSTRUCTION MEETING**

☐ a. **As outlined in Checklist 1: Design Considerations Common to All Permeable Pavements Summary**—Confirm that specifications are clear and review each of the items listed below with emphasis in materials testing, avoiding unspecified soil compaction to the subgrade and proper installation of erosion and sediment control per Best Management Practices.

☐ **3. SITE INSPECTION PRIOR TO INSTALLATION**

☐ a. **Site walk**—Walk through the site with project engineer, geotechnical engineer and builder/contractor/subcontractor to review erosion and sediment control plan/stormwater pollution prevention plan (SWPPP).

☐ b. **Construction sequence**—Determine when the permeable pavement is to be built in project construction sequence. Determine measures for protection and/or surface cleaning, if needed depending on sequence.

☐ c. **Aggregate storage**—Identify storage location for aggregate material, which typically includes identifying an impervious surface or an area covered with a geotextile to ensure no soils/fines enter the aggregate during storage.

☐ d. **Access routes**—Identify access routes for delivery and construction vehicles, and ensure permeable pavement will have no or very little traffic.

☐ e. **Vehicle tire/track washing station**—Identify location and maintenance requirements (if specified in erosion and sediment plan/SWPPP)

☐ f. **Foundation walls**—Identify specifications and design for how the foundation is protected from water stored in the system (i.e., underdrain, liners, system below foundation elevation, etc.). Confirm with geotechnical engineer.

☐ g. **Water supply**—Confirm that the pavement system is at least 30 m (100 ft) from municipal water supply wells.

☐ **4. INSTALLATION**

Confirm the following:

☐ a. **Excavation**
- Utilities are located and marked by local service
- Excavated areas are marked with paint and/or stakes
- Excavation size and location conforms to plan
- Storage and disposal of excavated soil areas are defined, adhered to, and provide protection the of the permeable pavement placement location
- The subgrade has been prepared at the correct design elevation
- Keep wheeled vehicles off the pervious subgrade

Source: © VHB

CHECKLIST 2 ☑

Recommendations for Permeable Pavement Construction Procedures *(continued)*

4. INSTALLATION *(continued)*

☐ b. **Sediment management**

- Protect temporary soil stockpiles from run-on/runoff from adjacent areas and from wind erosion
- Ensure linear sediment barriers (if used) are properly installed, free of accumulated litter and built-up sediment less than ⅓ the height of the barrier.
- Ensure no runoff enters the permeable pavement system until soils are stabilized and contributing drainage areas are free of fines/sediment.

☐ c. **Soil subgrade**

- Ensure rocks, stumps, roots and debris are removed and voids refilled with open-graded aggregate.
- Confirm soil is compacted to specifications (if required) and field tested with density measurements per specifications.
- Confirm no groundwater seepage or standing water in system (if so, dewatering and possibly a dewatering permit may be required).
- Confirm slope and infiltration rate are suitable for infiltration.

☐ 5. GEOTEXTILE (if specified—typical on sides, not typical on bottom)

Confirm the following:

☐ a. **Meets specifications**

☐ b. **Placement and down slope overlap** if required—(Min. 0.6 m or 2 ft) common, but should conform to specifications and drawings

☐ c. **Sides of excavation covered**—With geotextile prior to placing aggregate base/subbase

☐ d. **No tears or holes**

☐ e. **No wrinkles**—Pulled taught and staked

☐ 6. IMPERMEABLE LINER (if specified)

Confirm the following:

☐ a. **Meets specifications**

☐ b. **Placement, shop or field welding and seals at pipe penetrations**—Meets specifications

☐ 7. DRAIN PIPES/OBSERVATION WELLS

Confirm the following:

☐ a. **Size, perforations, locations, slope and outfalls**—Meets specifications and drawings

☐ b. **Verify elevation of overflow pipes**

☐ 8. SUBBASE, BASE, BEDDING AND JOINTING AGGREGATES

Confirm the following:

☐ a. **Sieve analysis**—From quarry conforms to specifications

☐ b. **Spread (not dumped)**—With a front-end loader to avoid aggregate segregation

Source: © VHB

CHECKLIST 2 ☑

Recommendations for Permeable Pavement Construction Procedures *(continued)*

☐ 9. EDGE RESTAINTS

Confirm the following:

☐ a. **Elevation, placement and materials**—Meets specifications and drawings

☐ 10. PERMEABLE PAVEMENT SURFACES

☐ a. **Meet ASTM standards or test methods**—As specified per manufacturer's test results

☐ b. **Elevation, slope, and placement/compaction**—Meets drawings and specifications

☐ c. **PICP**—No cut paver subject to tire traffic less than ⅓ of a whole paver; all pavers within 2 m (6.5 ft) of the laying face fully compacted at the completion of each day. Laying pattern and joint widths to meet drawings and specifications

☐ d. **Surface tolerance of completed permeable pavement surface**—Deviates no more than ±10 mm (0.4 in.) under a 3 m (10 ft) long straightedge

☐ FINAL INSPECTION

Confirm the following:

☐ a. **Surface swept clean**

☐ b. **Elevations and slope(s)**—Conforms to drawings

☐ c. **Transitions to impervious paved areas**—Separated with edge restraints or seams per details as required

☐ d. **Surface elevation of pavement surfaces**—3 to 6 mm (0.1 to 0.2 in.) above adjacent drainage inlets, concrete collars or channels to allow for minor settlement if specified

☐ e. **For PICP**—Bond lines for paver courses: ±15 mm (±0.6 in.) over a 15 m (50 ft) string line

☐ f. **Stabilization of soil in area draining into permeable pavement**—Min. 6 m (20 ft) wide vegetative strips if allowed.

☐ g. **Emergency overflow systems installed per specifications**—Confirm that any overflows will be properly conveyed.

☐ h. **No run-on from non-vegetated soil or unstable slopes to permeable pavement**

☐ i. **Test surface for infiltration rate per specifications**—Recommend using ASTM C1701-09 for pervious concrete and porous asphalt and ASTM 1781-13 for PICP.

*This information is a suggested framework for a checklist and should not be considered an exhaustive or complete list of all items that should be reviewed. Advice and direction from a competent professional in the field should be sought for site specific application of any and all material included in this report.

Source: © VHB

1.14 References

American Association of State Highway and Transportation Officials (AASHTO) (1993). "Guide for Design of Pavement Structures." American Association of State Highway and Transportation Officials, Washington, DC.

American Association of State Highway and Transportation Officials (AASHTO) (2004). "Guide for Mechanistic-Empirical Design of New And Rehabilitated Pavement Structures." National Cooperative Highway Research Program, Transportation Research Board, National Research Council, American Association of State Highway and Transportation Officials, Washington, DC.

Ballestero, T. N., Uribe, F., Roseen, R., and Houle, J. (2011). "The Porous Pavement Curve Number." *Proc., Greening the Environment, National Low Impact Development Conference*, Philadelphia, PA; NC State University, Raleigh, NC.

Bean, E. Z., Hunt, W. F., and Bidelspach, D. A. (2005). "A Monitoring Field Study of Permeable Pavement Sites in North Carolina." *Proc., 8th Biennial Conference on Stormwater Research & Watershed Management*, Southwest Florida Water Management District, Tampa, FL.

Bean, E. Z., Hunt, W. F., and Bidelspach, D. A. (2007). "Field Survey of Permeable Pavement Surface Infiltration Rates." *Journal of Irrigation and Drainage Engineering*, 133(3), 247–255.

Brattebo, B. O., and Booth D. B. (2003). "Long-term Stormwater Quantity and Quality Performance of Permeable Pavement Systems." *Water Resources*, Elsevier Press, 37, 4369–4376.

Cedergren, H. R. (1989). "Seepage, Drainage and Flow Nets." 3rd Ed., Wiley Interscience, New York.

Chang, N., Editor. (2010). *Effects of Urbanization on Groundwater: An Engineering Case-Based Approach for Sustainable Development.* American Society of Civil Engineers E-book, Reston, VA.

Chopra, M., and Wanielista, M. (2007). "Report 2 of 4: Construction and Maintenance Assessment of Pervious Concrete Pavements" *Performance Assessment of Portland Cement Pervious Pavement,* Stormwater Management Academy, University of Central Florida, Orlando, FL, R. Browne, ed., final report FDOT Project BD521-02,<http://www.dot.state.fl.us/research-center/completed_proj/summary_rd/fdot_bd521_02_rpt2.pdf>.

Colandini, V., Legret, M., Brosseaud, Y., and Baladès, J.D. (1995). "Metallic Pollution in Clogging Materials of Urban Porous Pavements." *Water Science and Technology*, 32(1), 57–62.

Collins, K. A., Hunt, W. F., and Hathaway, J. M. (2008). "Hydrologic Comparison of Four Types of Permeable Pavement and Standard Asphalt in Eastern North Carolina." *Journal of Hydrologic Engineering*. 13(12).

Day, G. E., Smith, D. R. and Bowers, J. (1981). "Runoff and Pollutant Abatement Characteristics of Concrete Grid Pavements." *Bulletin 135*, Virginia Water Resources Research Center, Virginia Tech, Blacksburg, Virginia.

Dierkes, C., Kuhlmann, L., Kandasamy, J., and Angelis, G. (2002). "Pollution Retention Capability and Maintenance of Permeable Pavements." *Global Solutions for Urban Drainage, 9th International Conference on Urban Drainage,* Portland, OR.

Fassman, E.A., Blackbourn, S. (2011). Runoff Water Quality Mitigation by Permeable Pavement. *Journal of Irrigation and Drainage*. 137(11): 720–729.

Gilbert, J.K. and Clausen, C. C., (2006). "Stormwater Runoff Quality and Quantity from Asphalt, Paver, and Crushed Stone Driveways in Connecticut." *Water Research*, 40, 826–832.

Hawkins, R. H., Ward, T. J., Woodward, D. E., and Van Mullem, J. A. (2009). "Curve Number Hydrology—State of the Practice." American Society of Civil Engineers (ASCE), Reston, Virginia.

Houle, J., Roseen, R., and Ballestero, T. (2010). *UNH Stormwater Center 2009 Annual Report.*. University of New Hampshire Stormwater Center, Cooperative Institute for Coastal and Estuarine Environmental Technology, Durham, NH.

Houle, K. M., Roseen, R. M., Ballestero, T. P., Houle, J. J. (2009). "Performance Comparison of Porous Asphalt and Pervious Concrete Pavements in Northern Climates," *Proc., Stormcon 2009*, Forster Publications, Santa Monica, CA.

Hunt, W. F. and Collins, K. (2008). "Permeable Pavement: Research Update and Design Applications." Department of Biological and Agricultural Engineering, Cooperative Extension Service, North Carolina State University, Raleigh, NC, AGW 588 E08 50327, 07/08/BS.

Iowa Stormwater Management Manual (2009). Porous Asphalt Pavement, Section 1J-3 - Porous HMA Pavement, Iowa Department of Natural Resources.

James, W., and Thompson, M. K. (1994). "Provision of Parking-Lot Pavements for Surface Water Pollution Control Studies." *Proc., Stormwater and Water Quality Management Modeling Conference*, Toronto, Ontario.

James, W. and Verspagen, B. (1996). "Thermal Enrichment Of Stormwater By Urban Pavement." *Proc., Stormwater and Water Quality Management Modeling Conference*, Toronto, Ontario.

James, W., and Thompson, M. K. (1996). "Contaminants from Four New Pervious and Impervious Pavements in a Parking Lot." *Proc., Stormwater and Water Quality Management Modeling Conference*, Toronto, Ontario.

James, W., and Kresin, C. (1996). "Observations of Infiltration Through Clogged Porous Concrete Block Pavers." *Proc., Stormwater and Water Quality Management Modeling Conference*, Toronto, Ontario.

James, W., and Gerrits, C. (2002). "Maintenance of Infiltration Rates in Modular Interlocking Concrete Pavers with External Drainage Cells", *Proc., Stormwater and Urban Water Systems Modeling Conference*, Toronto, Ontario.

Karasawa, A., Toriiminami, K., Ezumi, N., and Kamaya, K. (2006). "Evaluation Of Performance Of Water-Retentive Concrete Block Pavements." *Proc., 8th International Conference on Concrete Block Paving*, Interlocking Concrete Pavement Institute Foundation for Education and Research, San Francisco, CA, Herndon, VA.

Kevern, J. T. and Zufelt, J. (2013). "Introduction to ASCE Monograph—Permeable Pavements in Cold Climates." *Proc., 10th International Symposium on Cold Regions Development* (CD-ROM), ASCE, Anchorage, AK, 1-13.

Legret, M., Nicollet, M., Miloda, P., Colandini, V., and Raimbault, G. (1999). "Simulation of Heavy Metal Pollution from Stormwater Infiltration through a Porous Pavement with Reservoir Structure." *Water Science Technology*, 39(2), 119–125.

North Carolina Department of Environment and Natural Resources (NCDENR) (2007a). "Section 18: Permeable Pavements." *Stormwater BMP Manual*, NCDENCR, Revised: June 01, 2012, Raleigh, NC.

North Carolina Department of Environment and Natural Resources (NCDENR) (2007b). "Section 18: Permeable Pavements." *Stormwater BMP Manual*, NCDENCR, Revised: Sept. 01, 2010, Raleigh, NC.

Ohio Department of Natural Resources, Division of Soil and Water Conservation (2006). "Chapter 2: Post Construction Stormwater Management Practices." *Rainwater and Land Development Manual*, Ohio's Standards for Stormwater Management Land Development and Urban Stream Protection, 3rd Ed., Columbus, Ohio, <http://www.dnr.state.oh.us/tabid/9186/default.aspx>.

Pagotto, C, Legret, M., and Le Cloirec, P. (2000). "Comparison of the Hydraulic Behaviour and the Quality of Highway Runoff Water According to the Type of Pavement." *Water Research*, 34(18): 4446-4454.

Pratt, C. J., Mantle, D. G., and Schofield, P. A. (1995). "UK Research into the Performance of Permeable Pavement: Reservoir Structures in Controlling Stormwater Discharge Quantity and Quality." *Water Science Technology*, 32(1), 63–69.

Pratt, C. J., Newman, A. P. and Bond P.C. (1999). "Mineral Oil Biodegradation within a Permeable Pavement: Long Term Observations." *Water Science Technology*, Elsevier Science Ltd., Great Britain, 39(2), 103–109.

Roseen, R. M., Ballestero, T. P., Houle, K. M., Heath, D., and Houle, J. J. (2013-Accepted). "Assessment of Winter Maintenance of Porous Asphalt and Its Function for Chloride Source Control." *Journal of Transportation Engineering*.

Rowe, A., Borst, M., and O'Connor, T. (2010). "Chapter 13: Environmental Effects of Pervious Pavement as a Low Impact Development Installation in Urban Regions." *The Effects of Urbanization on Groundwater: An Engineering Case- based Approach for Sustainable Development*. N. B. Chang, ed., ASCE Publications.

Rhode Island Department of Environmental Management and Coastal Resources Management Council (RI CRMC) (2010). *Rhode Island Stormwater Design and Installation Standards Manual*, Providence, RI.

Rushton, B.T. (2001). "Low-Impact Parking Lot Design Reduces Runoff and Pollutant Loads." *Journal of Water Resources Planning and Management*, 127(3), Paper No. 22349, American Society of Civil Engineers, Reston, VA.

Smith, D.R. (2011). "Permeable Interlocking Concrete Pavements." Interlocking Concrete Pavement Institute, 4th Ed., Herndon, VA.

Swan, D. J. and Smith, D. R. (2009). "Development of the Permeable Design Pro Permeable Interlocking Concrete Pavement Design System." *Proc., 9th International Conference on Concrete Block Paving*, Argentina Concrete Block Association, Buenos Aires, Argentina.

Vancura, M., Khazanovich, L., and MacDonald, K. (2010). "Performance Evaluation of In-Service Pervious Concrete Pavements in Cold Weather." Department of Civil Engineering, University of Minnesota, Minneapolis, MN.

Van Seters, T. (2007). "Performance Evaluation of Permeable Pavement and a Bioretention Swale." *Interim Report #3*, Seneca College, King City, Ontario, Toronto, and Region Conservation Authority, Downsview, Ontario.

Virginia DCR Stormwater (2011). "Permeable Pavement." *Design Specification No. 7*, Version 1.8, <http://chesapeakestormwater.net/2012/03/design-specification-no-7-permeable-pavement> (June 27, 2013).

Porous Asphalt Fact Sheet

DESCRIPTION

Porous asphalt pavements include a permeable asphalt surface underlain by an open-graded aggregate choker course and a reservoir bed. Porous asphalt systems allow for stormwater filtration/infiltration and storage as well as a structural pavement in a single system. The bed depth is based on structural load, desired storage and frost depth requirements. Permeable pavement systems are usually placed on uncompacted subgrade to facilitate infiltration, but may include an underdrain and liner if necessary.

Source: CH2M Hill

APPLICATIONS

POTENTIAL APPLICATION		NOTES
Overflow Parking	Yes	
Primary Parking Areas (most heavily used)	Yes	Applicable with high durability mixes only; proper base/subbase design for loads
Sidewalks/Pathways	Yes	
Drive/Aisles	Yes	Must design for expected loads and use of parking lot; not recommended for routine truck traffic
Roads/Highways	Limited	Only with proper load and durability design; heavy duty pavement will be required*
Access Drives/Ring Roads	No (for heavy traffic)	With proper load and durability design, could be used for lighter traffic; heavy duty pavement expected to be required*
Loading Areas	No	Porous asphalt inappropriate due to spill concerns
Frequent Truck Traffic	Limited	Only with proper load and durability design considerations; heavy duty pavement will be required*

*Availability of heavy duty porous asphalt may be limited. Low to moderate durability porous asphalt is more commonly available.

Porous Asphalt Fact Sheet continued

BENEFITS

- Very high pavement surface permeability from 430 cm/hr to more than 1,250 cm/hr (170 in./hr to more than 500 in./hr); Reduces or eliminates runoff from pavement

- Provides groundwater recharge and volume reduction, peak rate control (increased with basecourse storage), and water quality treatment

- High level of pollutant removal; pollutant removal is greater at lower infiltration rates.

STORMWATER QUANTITY FUNCTIONS		NOTES
Volume	High	Volume control up to 100% with total infiltration; may mitigate for greater area with additional storage
Groundwater Recharge	High	Recharge high if soils allow; raised outlet above bottom of storage will increase recharge where under drains are present
Peak Rate	Medium/High	Peak discharge rate control increased with increased storage and/or high permeability in existing soils
STORMWATER QUALITY FUNCTIONS**		NOTES
Total Suspended Solids	High	Must vacuum to minimize sediment on pavement; must prevent soils or other materials flowing on to pavement
Total Nitrogen/ Total Phosphorus	Med/High	Total nitrogen removal varies; moderate 40% to 60% total phosphorus removal
Metals	High	Removal greater than 95% has been reported
Temperature	High	Water discharges to ground

** Water quality functionality is dependent on optional use of filter course; infiltration systems load reduction is determined by volume reduction and treatment efficiency

SITE CONSTRAINTS/CONCERNS

- Water table/bedrock separation: 0.61 m (2 ft) minimum

- Soils: HSG A and B preferred; C and D may require underdrain

- Feasibility on steeper slopes: possible with design modifications to maintain internal grade control

- Areas with the potential for contamination sources: appropriate if liner and under drain provided

- Limit run-on from adjacent contributing surfaces to prevent clogging. Direct run-on from adjacent impervious areas should be limited to 1.5 times the porous asphalt surface area (the subsurface bed may have capacity to accept additional runoff via pipe or other means).

Porous Asphalt Fact Sheet continued

RECOMMENDED KEY DESIGN CRITERIA

- Site location soil permeability testing
- Uncompacted, permeable subgrade soils (for infiltration)
- Provide positive stormwater overflow from storage reservoir
- Base/Subbases and pavement structural design must suit application
- Avoid use in areas surrounded with fine soils or vegetation that could run onto or be deposited on the pavement.
- Should not be installed on wet aggregate or treated bases when air temperature is less than 13°C (55°F)
- Requires 48-hour curing time following installation (7 days recommended)

PERFORMANCE

GOOD—All season performance. In northern climates, the use of porous asphalt results in reduced salting requirements; however, no sand, cinders, or abrasives should be applied on or adjacent to the porous asphalt at any time.

OPERATION & MAINTENANCE

MEDIUM—Regular vacuum sweeping is required to maintain pavement surface porosity. Maintenance/vacuuming two (2) times per year as a minimum is recommended, but it generally depends on traffic intensity and adjacent soils/vegetation.

COSTS

MEDIUM—Costs are decreasing as demand for product increases. Check with the local asphalt batch plant on current prices and contingencies such as distance from batch plant and minimum batch sizes (tons).

REFERENCES

United States Environmental Protection Agency (2009). "Porous Asphalt Pavement." *National Pollutant Discharge Elimination System (NPDES)*, <http://water.epa.gov/polwaste/npdes/swbmp/Porous-Asphalt-Pavement.cfm> (Nov. 1, 2011).

2 Porous Asphalt and Permeable Friction Course Overlays

2.1 System Description

Porous asphalt typically consists of conventional warm mix asphalt (WMA) or hot mix asphalt (HMA) with significantly reduced fines resulting in an open-graded mixture that allows water to pass through an interconnected void space. The porous asphalt surface void space typically ranges from 18% to 25%. In comparison, voids for standard asphalt are typically 2% to 3%, and they are not interconnected.

The Franklin Institute in Philadelphia, PA conducted initial development of porous asphalt in the early 1970s. Some of these earlier installations are still functioning well (see **Figures I-1 and I-2**.) Significant mix design limitations were addressed in the early 2000s, resulting in durable long-lived pavements appropriate for a variety of uses.

Porous asphalt pavement is placed directly on an open-graded aggregate base (**Figure 2-1**) including a choker course or over an asphalt-treated permeable base (ATPB), but not on a conventional dense-graded base/subbase. Aggregate bases are comprised of clean angular stone. Rolling porous asphalt must be done carefully to provide a proper level of compaction that does not cause aggregate breakdown and/or result in excessive reduction of the surface porosity and pavement voids. Preparation of the aggregate subbase requires additional care to establish that sufficient compaction exists to provide structural support, while still maintaining sufficient infiltration capacity.

Porous asphalt

Choker course or asphalt treated permeable base

Open-graded base/subbase layers

Uncompacted soil subgrade

Figure 2-1
Typical porous asphalt system section
Source: © VHB

Because porous asphalt is essentially conventional WMA or HMA with reduced fines, it is similar in appearance to conventional asphalt pavement, although generally coarser in texture (**Figure 2-2**). Recent research in open-graded friction course (OGFC) for highway applications has led to additional durability improvements in porous asphalt through the use of additives (i.e., fibers, anti-stripping agents, etc.) and higher performance-grade binders. These should be considered for applications with increased load/traffic demands. Porous asphalt is suitable for use in most climates where conventional asphalt is appropriate. Many projects use both porous and dense-graded HMA/WMA to suit the project conditions and/or traffic loads.

Figure 2-2
Porous asphalt surface at Port of Portland, OR
Source: Century West Engineering

2.2 Applications—Site Constraints/Concerns

Porous asphalt is well-suited for primarily low to moderate traffic parking areas, recreational areas (e.g., basketball and tennis courts), driveways, and low-speed/low-volume roadways (see **Figures 2-3 through 2-8** for project examples). Porous asphalt can also be used for sidewalks and pathways. Higher volume roads and highways have also been constructed of porous asphalt with some success (e.g., see Transportation Research Board, 1992 and **Figure 2-7**). However, porous asphalt has generally not used in these types of applications because of the required earthwork (cutting and filling), grading, sequencing and other issues typically involved with highway construction.

For stormwater infiltration, porous asphalt pavement design must consider site conditions that affect the infiltration process. While still presenting design challenges, even low permeability soils can make for effective installations and these soils often have higher pollutant removal capacities (e.g., cation exchange capacities). For example, a soil with a permeability of only 0.13 cm/hr (0.05 in./hr) can infiltrate over 3.1 cm (1.2 in.) of runoff in less than one day.

Figure 2-3
Porous asphalt residential lane, Pelham, NH
Source: Roseen, R. M.

Figure 2-4
Parking lot with standard aisle and porous asphalt stalls, Morris Arboretum, Philadelphia, PA
Source: CH2M HILL

Figure 2-5
Porous asphalt path, Grey Towers National Historic Site, PA
Source: CH2M HILL

Figure 2-6
Commercial parking lot with standard pavement and porous asphalt, Greenland Meadows, Greenland, NH
Source: Roseen, R. M.

Figure 2-7
Permeable pavement test section on state highway using open-graded friction course over and asphalt treated permeable base (ATPB), MaineDOT, Portland, MA
Source: Roseen, R. M.

Figure 2-8
Porous asphalt basketball court, Upper Darby, PA
Source: CH2M HILL

General guidelines for designing stormwater infiltration in porous asphalt systems (functionally, the sub-bases of the porous asphalt system) include items outlined in **Chapter 1** plus the following:

Climate Considerations

While porous asphalt is generally appropriate for climatic regions that use conventional asphalt, several additional factors must be considered for design in areas with prolonged freezing temperatures. These factors include frost depth, frost durability of the materials, frost heave of the subgrade, and frost durability of the saturated system. Installation of appropriate drainage elements and a capillary barrier, such as a layer of open-graded stone to prevent water wicking into the base from below (if using a sand layer), are needed to prevent extended saturation of the base and subbase layers resulting in the potential for freeze-thaw issues. If these issues are not adequately addressed, frost heave of a saturated subgrade may cause excessive movement during long periods of freezing weather and may result in significant loss of subgrade support during spring thaws.

While these factors should be considered, porous asphalt pavements have been known to perform well in cold climates most likely because of their permeable and open-graded components, allowing for expansion of freezing water. However, water within the system should never be stored in the pavement surface itself, especially when freezing conditions are possible. Studies at the University of New Hampshire Stormwater Center (UNHSC) and in northern Europe have shown that surface infiltration rates are maintained despite a frozen subsurface. This is due to the open-graded materials and a reduced duration of frost within porous asphalt (Roseen et al 2012).

Porous asphalt pavements have been found to be less susceptible to damage as a result of freeze and thaw cycles than conventional pavements (Backström 1999). In somewhat frost-susceptible soils, increasing the minimum base depth to 35 to 55 cm (14 to 22 in.) may be warranted depending on loading and specific soil conditions. In extremely frost susceptible soils, the base/subbase and/or improved soils can be placed down to the full frost depth. In areas with substantial frost penetration, UNHSC recommends that the bottom of the stormwater base/subbase beneath pervious pavement be extended at least 65% of the local design frost depth (UNHSC 2009).

Additionally, porous asphalt is less likely to form black ice, often requiring less plowing and fewer deicing chemicals. This is mostly due to the dark color of the pavement and high porosity. While thin layers of snow are melted on sunnier days or by the heat of the vehicle traffic, the melted ice water enters the pores of the pavement. Water therefore should not remain on the surface and be susceptible to refreezing, eliminating or reducing the need for repetitive deicer applications. UNHSC found that a porous asphalt parking lot reduced the need for winter maintenance salt by approximately 50% to 75% (Roseen and Ballestero 2008; UNHSC 2009). The reduction in the amount of deicing materials is primarily observed with respect to the reduced development of black ice. Reductions will vary by application based on site-specific conditions including sun exposure, slope, traffic, etc.. As shown in **Figure 2-9**, the standard pavement is covered with snowmelt that is likely to form black ice at night. However, it should be noted that porous asphalt pavements can be more challenging to de-ice if compacted snow and ice develop.

Figure 2-9
Dry porous asphalt in foreground and conventional pavement in the background, Cayuga Medical Center, Ithaca, NY
Source: CH2M HILL

Abrasives, such as sand or cinders for winter traction, should not be applied on or near the porous asphalt pavement. It is recommended that abrasives not be used at all on sites with porous asphalt to minimize sand being tracked onto the pavement by vehicles and to prevent accidental applications to the porous asphalt surfaces. Frequent snow plowing is an essential component, but should be done carefully to avoid damaging the pavement surface.

In warmer climates, clogging of several porous asphalt systems has been anecdotally attributed to long-term draindown of asphalt binder in the pavement layer (Ferguson 2005). The asphalt binder is thought to slowly drain down during hot periods, mixing with sediment, dust from pavement wear, etc. to form a clogging layer in the pavement. According to the National Asphalt Pavement Association (NAPA), this is more of a concern during construction when the WMA/HMA mix is at its highest temperature. Careful mix design—especially the selection of the asphalt binder grade, mix production temperatures, and the use of additives—is recommended to reduce of the risk of draindown. The addition of fibers to the porous asphalt mix has been shown to effectively manage draindown and contribute to material strength. Fibers are added at the asphalt patch plants, but must be coordinated with the producer and detailed in the specifications. Additional research on the draindown phenomenon is also warranted.

2.3 General Design Criteria—General Specifications

General Design Considerations

If intended to achieve significant stormwater infiltration, porous asphalt pavement systems should follow general design guidelines in Chapter 1 and the following:

Structural Design Considerations

Most porous asphalt systems are intended for relatively light automobile traffic and do not require significant structural design modifications (NAPA 2008). In situations where the porous asphalt will need to support truck traffic, it is critical to consider the structural requirements of the pavement system and make appropriate design modifications for the anticipated heavy loads. In many cases, the runoff volume and soil infiltration rate will control the thickness of the base/subbase layers, while traffic loads will control the thickness of the porous asphalt surface and asphalt treated permeable base (ATPB). While the vast majority of porous asphalt pavement systems have been designed for automobile traffic with only occasional trucks (e.g., delivery trucks), limited case studies have shown positive experience supporting the use of porous asphalt systems for trucks. At a site in Salem, OR, the ATPB courses withstood heavy construction traffic, probably because of their thickness and due to the fact that—unlike conventional bases—they are as strong when saturated as when dry (NAPA 2008). Table 2-1 provides recommended AASHTO layer coefficients for the structural evaluation of porous asphalt and Table 2-2 provides recommended minimum thicknesses for the porous asphalt surface for different traffic loadings.

Hydrologic Design

The hydrologic design of porous asphalt is similar to that of other types of permeable pavements. See Chapter 1 and Chapter 9.

Porous Asphalt Materials

Typical porous asphalt mix characteristics are listed in Table 2-3. Guidance for preparing specifications has been published by UNHSC (2009) and others. State asphalt associations may also provide additional guidance. All specification guidelines and information should be evaluated on a case-by-case basis and will vary due to local materials, practices, and the specifics of the project (i.e., size, traffic, climate, etc.).

The thickness of the porous asphalt layer depends on the structural design requirements, but is typically 6.4 to 10 cm (2.5 to 4 in.) or minimum of 3.8 cm (1.5 in.) per layer for two-layer installations, where one layer is

Table 2-1 Recommended AASHTO Layer Coefficients for Porous Asphalt

MATERIAL	LAYER COEFFICIENT
Porous Asphalt	0.35–0.42
Asphalt Treated Permeable Base (ATPB)	0.30–0.35
Open-Graded Aggregate Base Layers	0.10–0.14

Source: NAPA 2008

Table 2-2 Minimum Compacted Pavement (combined porous asphalt and ATPB, if used) Thickness

TRAFFIC LOADING	MINIMUM COMPACTED THICKNESS cm (in.)
Parking—few or no trucks	7 (2.75)
Residential Street—some trucks	10 (4.00)
Heavy Trucks	15 (6.00)

Source: NAPA 2008

Table 2-3 Typical Porous Asphalt Material Characteristics

AGGREGATES	
SIEVE SIZE, mm (in.)	**PERCENT PASSING (%)**
19 (0.75)	100
12.5 (0.50)	85–100
9.5 (0.375)	55–75
4.75 (No. 4)	10–25
2.36 (No. 8)	5–10
0.075 (No.200)	2–4
BINDER AND MIX ADDITIVES	
Binder Content (AASHTO T164)	6–6.5
Fiber Content By Total Mixture Mass	0.3 cellulose or 0.4 mineral
Rubber Solids (SBR) Content By Weight of the Bitumen	1.5–3 or TBD
PHYSICAL CHARACTERISTICS/TESTING	
Air Void Content (ASTM D6752/AASHTO T275)	18.0–22.0
Draindown (ASTM D6390)	≤ 0.3
Retained Tensile Strength (AASHTO 283)*	≥ 80
Cantabro Abrasion Test on Unaged Samples (ASTM D7064-04)**	≤ 20
Cantabro Abrasion Test (ASTM D7064-04) on 7-Day Aged Samples**	≥ 30

* AASHTO 283 is sometimes difficult to run on porous mixes—see NAPA guidance (NAPA 2008) for more information.
** As described in ASTM D7064-08
Source: Roseen, R. M., Ballestero, T. P., Briggs, J. F., and Pochily, J. (2009). "UNHSC Design Specifications for Porous Asphalt Pavement and Infiltration Beds." University of New Hampshire Stormwater Center, Durham, NH

porous asphalt and one layer is ATPB. Greater thickness will be required for heavier traffic loads. Typical permeability rates for new, properly constructed porous asphalt have been estimated to be from 430 cm/hr to more than 1,250 cm/hr (170 in./hr to more than 500 in./hr) (summarized from multiple sources in Ferguson 2005). The permeability of the pavement surface is expected to decrease somewhat over time depending on traffic, sediment/debris loading, and the level of maintenance.

Although the base layers under porous asphalt often have the capacity to store runoff from adjacent impervious areas, direct surface sheet flow of runoff from impervious areas (i.e., conventional pavements) to the pervious pavement surface should be minimized; a maximum ratio of approximately 1.5:1 is recommended. High sheet flow loading to porous asphalt may lead to premature clogging or could require excessive maintenance of the porous asphalt.

Asphalt-Treated Permeable Base (ATPB) (optional)

In a two-layer porous asphalt system, the asphalt-treated permeable base is a coarse layer directly beneath the porous asphalt. Compared to open-graded aggregate, this layer acts as a much more stable base for installing the porous asphalt surface. ATPB typically consists of relatively coarse aggregate with 3% to 4% liquid asphalt binder and can be used to add structural strength to the pavement. Many state DOTs have ATPB specifications that may be acceptable for a porous asphalt system. This base layer, which is typically 7.6 to 15.2 cm (3 to 6 in.) in depth, also provides a more stable paving platform, which may help achieve a better finished surface.

Choker Course (optional)

A choker base course (often ASTM No. 57) aggregate between the porous asphalt surface and the underlying reservoir course is typically included in order to provide an even surface for paving and to prevent excessive pavement loss into the reservoir coarse aggregate. Recent experience has indicated that some contractors are moving away from using the choker course because it may not fully achieve its primary purpose—to provide a more stable paving surface. For single lift paving, a choker layer should be considered. When an ATPB is used, a choker layer is not required (**Figure 2-10**). If a choker course is used, the aggregate should be sized to interlock with the aggregate in the underlying reservoir layer.

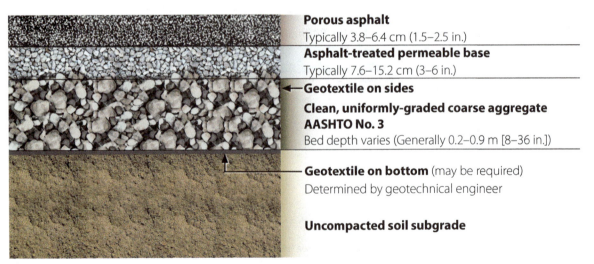

Porous asphalt
Typically 3.8–6.4 cm (1.5–2.5 in.)

Asphalt-treated permeable base
Typically 7.6–15.2 cm (3–6 in.)

Geotextile on sides

Clean, uniformly-graded coarse aggregate
AASHTO No. 3
Bed depth varies (Generally 0.2–0.9 m [8–36 in.])

Geotextile on bottom (may be required)
Determined by geotechnical engineer

Uncompacted soil subgrade

Figure 2-10
Two-lift porous asphalt section without a choker course
Source: © VHB

Reservoir Course

The reservoir course should be composed of clean, uniformly graded aggregate (often ASTM No. 2 or No. 3) with approximately 30%–40% void space. Local aggregate availability typically dictates the specific aggregate gradation. The critical requirements are that the aggregate be uniformly graded, washed, durable, angular, and have a significant void content for water storage. The thickness of the reservoir course may be determined based on stormwater storage requirements, frost depth, structural requirements, depth to bedrock, seasonal high water table elevation considerations and site grading. The bed is typically 0.2 to 0.9 m (8 to 36 in.) deep. See Chapter 1 for more on the design of the reservoir course. Figure 2-11 shows a porous asphalt surface over a thin layer of choker course and the reservoir course.

Figure 2-11
Porous asphalt section, San Diego County Operations Center, CA
Source: CH2M HILL

Filter Course (optional)

A filter course of poorly graded sand is included in some designs for its additional water quality and hydrologic benefits. The filter course is located between the choker course and reservoir course at thicknesses of 20 to 30 cm (8 to 12 in.). In order to prevent the migration of the filter material (typically sand) into the reservoir layer, the filter course should be underlain by a choker course (under the sand layer) typically of 8 cm (3 in.) of pea gravel (Figure 2-12). Porous pavements systems with a filter course can provide a very high level of filtration prior to infiltration and provide exceptional lag times and water quality treatment.

Systems with solely course aggregate layers have more of an infiltration and sedimentation function. The fine gradation of the filter course is for enhanced filtration and delayed discharge, which can also be achieved with raised underdrains or raised outflow (see Figure 1-7 in Chapter 1). In some areas, the filter course thickness is factored in as the elevation from which to determine separation from groundwater.

The use of a filter course can provide exceptional lag times, often in excess of 1,000 minutes or more as reported (Roseen et al 2012). The lengthening of lag time and residence time within the pavement system provides two important distinct hydrologic benefits, including greater peak reduction and groundwater recharge volumes. By increasing the system lag (and residence) time, there is increased infiltration and recharge that is especially important for low hydraulic conductivity soils. Increase in recharge volume is achieved simply by the increase in residence time prior to draining. Great care is required to not over-compact materials as this can lead to a loss of infiltration capacity. The filter course material is sometimes referred to as bank run gravel, but may also be a crushing byproduct, typically with less than 2% passing the 0.075 mm (0.003 in.) (No. 200) sieve.

The filter course serves two important functions both for water quality as a filtration mechanism and for pavement structure as a load bearing element. Because of its functions, the filter course is a common quality control concern. If the filter course is over compacted, internal drainage will be affected resulting from a poorly drained aggregate subbase. This has the potential to affect the system longevity in cold climates by increasing susceptibility to frost heave. Conversely, an under-compacted filter course will result in a reduced load bearing capacity contributing to a reduction in pavement strength and durability. The construction and installation of a filter course is an important point to employ construction quality assurance. The quality assurance is commonly conducted by a third party or the supervising engineer. Filter compaction should be 90% to 95% standard proctor density measured by ASTM D698 and a hydraulic conductivity of 3 to18 m/day (10 to 60 ft/day) measured by D2434. Construction quality assurance should be conducted routinely and reported to the supervising engineer. The inclusion of quality control is a relatively inexpensive component of overall system construction costs and greatly reduces two common failure mechanisms. A project that decides to forgo the quality control is better served using a standard subbase, and omitting the filter course as the same concerns for infiltration capacity and compaction do not exist.

Filter course material should have a hydraulic conductivity of 3 to 18 m/day (10 to 60 ft/day) when compacted to a minimum of 95% standard proctor density unless otherwise approved. Great care is required to not over-compact materials as this can lead to a loss of infiltration capacity. The filter course

Porous pavement (combined porous asphalt and ATPB, if used)
7.5–15 cm (3–6 in.) as 1–2 lifts each 4–7.5 cm (1.5–3 in.)

Choker course
8–15 cm (3–6 in.) minimum ASTM No. 57

Filter course
Typically sand 20–30 cm (8–12 in.)

Choker course for filter course
Typically pea gravel for intermediate setting bed:
8 cm (3 in.) thickness of 1 cm (0.4 in.) stone

Reservoir course
10 cm (4 in.) minimum thickness of 2 cm (0.8 in.)
ASTM No. 2 or 3

Perforated underdrain 10–15 cm (4–6 in.) diameter
pipe with 5 cm (2 in.) minimum cover

Liner (may be required) for land uses where infiltration
is undesirable (e.g., hazardous materials handling,
contaminated soils, sole-source aquifer protection)

Uncompacted soil subgrade

Figure 2-12
Porous asphalt section with filter course and capillary barrier for frost protection
Source: UNH Stormwater Center as modified by VHB

material is sometimes referred to as bank run gravel, but may also be a crushing byproduct with less than 2% passing the 0.075mm (No. 200) sieve.

Geotextiles

Recommendations vary on the use of geotextiles in porous asphalt pavement systems. There have been limited reports that geotextiles have the potential to clog. At a minimum, geotextiles should be used on soils with poor load bearing capacity, high fines content, and always on the sides of the excavation to prevent in-migration of fines. All permeable pavements can be subject to soil movement and deposition, and they should carefully consider geotextile use (See **Figure 1-10**). Additional considerations for the use of geotextiles are described in **Chapter 1**. Geotextiles should conform to AASHTO M-288 Geotextile Specification for Highway Applications. Geotextile permeability should conservatively exceed the permeability of the subgrade by at least an order of magnitude.

Backup Inflow/Overflow Structures

Porous asphalt systems should have a backup method for water to enter the reservoir course in the event that the pavement becomes clogged or altered. In uncurbed lots, this backup drainage may consist of an unpaved, 30 to 60 cm (1 to 2 ft) wide, stone-filled edge drain connected directly to the bed (**Figure 2-13**). This ensures that the stormwater system as a whole will continue to function even if there is some surface runoff from the pavement surface.

Inlet structures can be used for cost-effective overflow control, and discharges can be directed to an existing storm drain system (where available). Catchbasin/manhole inlets for overflow or receiving

Figure 2-13
River stone edge offers a backup method for runoff to enter the reservoir course, Yeadon, PA
Source: CH2M HILL

discharge from underdrains may be designed with discharge control features similar to typical detention systems. These structures may vary in design depending on factors such as peak discharge rate and storage requirements. They should always include positive overflow from the system and flow to a location that will not create negative impacts.

2.4   Recommended Installation Guidelines

Qualified engineering and construction oversight are essential to porous asphalt construction because installation differs significantly from conventional pavements. Oversight includes a pre-construction meeting to review the construction process and discuss sequencing, staging, erosion, sedimentation control, and any project specific considerations that might influence the installation. In addition, submittal reviews and construction observations are strongly recommended for critical parts of the installation including review of the porous asphalt mix design, materials submittals, erosion, and sediment control provisions—ensuring that sediment is not tracked, washed, blown, or otherwise deposited in the system—base course layer construction and pavement installation. For systems intended to infiltrate, ensuring that soil subgrades are not compacted or over-compacted is critical. This requirement is often misunderstood by contractors. General construction sequencing recommendations are as follows:

1. Mobilization, site preparation, and the installation as well as maintenance of erosion and sediment control measures are performed first. When possible, the installation of porous asphalt systems and other infiltration systems should wait until the end of construction (**Figure 2-14**).

   Excavation to the soil subgrade elevation is the next step. The existing subgrade under the reservoir course should **not** be compacted or subject to construction equipment traffic prior to geotextile and stone aggregate installation. Excavators/backhoes should be used to excavate the bed area from adjacent locations, so that equipment is never running on exposed subgrade (**Figure 2-15**). Only very low ground pressure (<0.03 MPa or <4 psi) equipment is acceptable in the bed areas when excavation is within 0.3 m or 1 vertical foot of the final subgrade elevation. Where operation of equipment on subgrade is unavoidable and compaction occurs, remediation should be done by scarifying or tilling the subgrade to no less than 20 cm (8 in.) with a York rake, light tractor, or other means.

2. All bed bottoms should be as level as feasible to promote uniform infiltration. Systems specifically designed with a sloped subgrade surface should define size, distance, and materials used for berms or barriers if needed to prevent lateral flow and promote infiltration (see **Chapter 1, Section 1.5 Site Conditions**). This will prevent erosion within the reservoir layer on slopes and provide runoff storage capacity. See **Figure 1-13** in **Chapter 1**. Earthen berms, if used, should be left in place during excavation, if possible (**Figure 2-16**). These berms may not require compaction if proven stable during construction.

3. Geotextile and aggregate should be placed immediately after owner/engineer approval of the prepared subgrade (**Figure 2-17**). Geotextile, if used, should be placed in accordance with the manufacturer's standards and recommendations. Adjacent strips of geotextile should overlap a minimum of 40 cm (16 in.). The geotextile should also be secured at least 1.2 m (4 ft) outside the bed to help prevent any runoff or sediment from entering the storage bed. This edge strip should remain in place until all bare soils contiguous to the beds are fully stabilized and vegetated. As the site is fully stabilized, excess geotextile along bed edges can be cut back to the bed edge.

4. Clean, washed, open-graded aggregate (**Figure 2-18**) should be placed in the bed in 20 to 30 cm (8 to 12 in.) lifts and compacted to a maximum 95% and minimum 90% standard proctor density (ASTM D698/AASHTO T99). The density of the subbase courses are determined by AASHTO T191 (sand cone method), AASHTO T204 (drive cylinder method), AASHTO T238 (nuclear methods) or other approved alternates suitable for measuring the compaction of open-graded aggregate. When possible, construction equipment shall be kept off the bed bottom entirely and traffic on the aggregate should be minimized (**Figure 2-19**). Once the reservoir course aggregate is installed to the desired grade, either the filter course (if included in design) or choker course is installed. If used, install filter course aggregate in 20 cm (8 in.) maximum lifts to a minimum 95% standard proctor density (ASTM D698 or AASHTO T99). The density of the filter course should be verified by the engineer/inspector. The infiltration rate of the compacted filter course should be determined by ASTM D3385 or approved alternate and should be no less than 3 m/day (10 ft/day). A 2.5 to 5 cm (1 to 2 in.) layer of choker base course (typically ASTM No. 57) aggregate may be installed uniformly over the surface in order to provide an even surface for paving. A typical choker course depth is be 2 to 3 times the thickness of the maximum aggregate size. As previously noted, recent experience indicates that some contractors are moving away from using the choker course because it may not be fully achieving its intended purpose, which is to provide a more stable paving surface. For a single layer surface, the choker course should be included. For a two-layer surface (porous asphalt over ATPB) a choker course is not typically necessary. If a choker course is used, the aggregate should be sized to interlock with the aggregate in the storage bed.

Figure 2-14
Porous asphalt installation, Peabody, MA
Source: VHB

Figure 2-15
Excavator operating from outside the excavation to prevent compaction of subgrade, Modesto, CA
Source: CH2M HILL

Figure 2-16
Earthen berms and geotextile, San Diego, CA
Source: CH2M HILL

Figure 2-17
Geotextile being placed by hand to avoid compaction of subgrade, Modesto, CA
Source: CH2M HILL

Figure 2-18
Clean, washed, crushed storage bed aggregate
Source: CH2M HILL

Figure 2-19
Aggregate placed in the bed from the edge to prevent compaction, Chapel Hill, NC
Source: CH2M HILL

Figure 2-20
Installation of porous asphalt over the aggregate base layer with a tracked paver, Warrington, PA
Source: CH2M HILL

Figure 2-21
A small roller smooths seams of porous asphalt surface course, Yeadon, PA
Source: CH2M HILL

5. Porous asphalt should not be installed on wet aggregate, treated bases, or when the ambient air temperature is below 13°C (55°F). The production temperature of the bituminous mix should be determined by the results of draindown testing (ASTM D6390) and recommendations of the asphalt supplier, but typically ranges between 135°C and 150°C (275°F and 302°F). Porous asphalt is typically placed in one lift for small projects or two lifts for thicker sections. The two-layer pavement/ATPB surface is installed directly over the aggregate base layers to the specified finish thickness (**Figure 2-20**). Two to three passes with an 8 to 10 ton static steel wheel roller is required for proper compaction (i.e., air voids of 18% to 22%). Additional rolling could reduce surface course porosity and/or cause aggregate breakdown. Additional rolling with a small roller to smooth seams and remove marks is normally required (**Figure 2-21**). Rollers should move slowly and uniformly to prevent displacement of the mix, and they should not be stopped or parked on the freshly placed mat.

6. In a two-lift application (porous asphalt surface pavement over ATPB), site development activities may need to occur prior to the final lift of paving, but after construction of the pavement system has begun. In those cases, paving the base layer of ATPB to temporarily accommodate construction traffic has been successful (NAPA 2008). The base layer is covered and protected from sediment and other deleterious material with geotextile. Care must be taken to maintain the geotextile during construction so that debris, soil, etc., do not contaminate the ATPB layer, which could clog the voids. Once construction activities are complete, the geotextile is removed and the base pavement layer vacuumed and flushed with water prior to the final lift of porous asphalt paving.

2.5 Post-construction Operation and Maintenance

After the porous asphalt installation, no vehicular traffic of any kind should be permitted on the pavement surface until cooling and hardening or curing has taken place, and in no case within the first 48 hours (7 days minimum is recommended).

Figure 2-22
Sediment-laden stormwater runoff clogs the porous asphalt, Campbelltown, PA
Source: CH2M HILL

The primary goal of porous asphalt system maintenance is to prevent the pavement surface and underlying stormwater bed from becoming clogged by sediment and/or debris. **Figure 2-22** shows a project site where erosion and sediment controls were removed too soon, causing sediment-laden stormwater runoff to temporarily clog the porous asphalt. The porous pavement surface should typically be thoroughly vacuumed twice per year with a commercial street cleaning unit, although the optimal frequency is site specific, depending on traffic intensity and adjacent soils/vegetation.

2.6   Cost Information

Much of the increased unit cost of porous asphalt pavement systems comes from the aggregate stormwater bed, which is generally thicker than the aggregate base for conventional pavement. However, this additional cost is often offset by the significant reduction in the required conventional stormwater management elements like inlets, pipe, basins, right of way, detention ponds, curb, and gutter etc. Also, because pervious pavements are often incorporated into the natural topography of a site, less earthwork may be required. Porous pavements with reservoir layers often eliminate the need and associated costs of detention basins or other similar stormwater management systems. When these factors are considered, porous pavement with infiltration can be approximately the same cost or often even less expensive than conventional pavement with its associated stormwater management facilities. See **Table B-4** in **Appendix B** for permeable pavement cost range estimates.

The cost for porous asphalt surfaces are generally 20% to 50% higher than conventional asphalt on a unit area basis (not including the associated required stormwater facilities). This is due to the use of additives (e.g., fibers, polymers) to control draindown and increase stability. Costs for porous asphalt (not including the reservoir course) range from about $21 to $38/m^2 ($2 to $3.50/sf). As a complete system, the estimated cost is $65 to $130/m^2 ($6 to $12/sf). As expected, thicker and more complex sections (multiple layers) are typically more expensive.

In some instances for new development, total project cost savings have been observed when using porous pavements through the avoidance of additional stormwater infrastructure. For example, a 6% saving was realized for a porous asphalt residential road application by avoiding the use of curbing, pipe, catch-basins, detention basins, and outlet control structures (Gunderson 2010).

2.7   Stormwater Benefits

Volume Reduction

Porous pavements with infiltration provide an excellent means of capturing and infiltrating runoff and the discharge resembles shallow depth groundwater drainage, as is the goal for low impact development designs. The reservoir course below the pavement provides storage for runoff during storm events, while the uncompacted subgrade allows infiltration of runoff into the underlying soils. Infiltration and runoff reduction is possible for a range of soil types, including low conductivity soils. A runoff volume reduction of 25% was observed for an underdrained porous pavement with hydrologic group C soils, while 92% was measured for a site with hydrologic group B soils (UNHSC 2009).

Pollutant Removal

Porous asphalt systems are effective in reducing pollutants such as total suspended solids (TSS), metals, phosphorus, oil, and grease. The porous asphalt section, the filter course (if used), the stormwater bed, and the underlying soil each provide pollutant removal. A porous asphalt study conducted by the UNHSC showed more than 95% reduction for total suspended solids, total petroleum hydrocarbons, and total zinc as well as a 25% removal for total phosphorus, but no significant treatment for nitrogen. Additionally, a study in Texas examined the quality of runoff from a conventional asphalt pavement and an open-graded friction course (OGFC). This study indicated that even the thin porous asphalt layer above the conventional highway pavement removes a significant amount of the pollutants normally associated with runoff from pavement (Barrett 2007).

Peak Rate Control

Properly designed pervious pavement systems can provide effective management of peak runoff rates from storm events. The stormwater bed below the pervious pavement acts as a storage reservoir during large storm events, even while runoff infiltrates into the underlying soils. Outlet structures can be designed to manage peak rates with the use of outlet controls, and carefully designed systems can manage peak rates for storms up to and including the 100-year storm.

See **Appendix B** for a summary of stormwater benefits, costs, and hydrology.

2.8   Permeable Friction Course Overlays

Another method of porous asphalt use is on top course over standard impermeable asphalt. This is called a permeable friction course (PFC) overlay **(Figure 2-23)**. Since the 1970s, permeable (or porous) friction course overlays, also known as open-graded friction courses (OGFC), have been installed on selected roadways in an effort to make them quieter and safer. The pervious course leads to shorter stopping distances for cars, quicker surface drying periods, less splash and spray during precipitation as well as reduction of noise.

Stormwater runoff travels through the voids in the PFC overlay until it reaches the boundary of the PFC and underlying impermeable asphalt. The water then flows along this convex boundary layer toward the shoulders of the road where it discharges along the road perimeter. Since stormwater runoff has a route of discharge, it does not pond on the road surface, which can increase vehicle skid resistance. Impermeable roadways pose a greater risk of hydroplaning, in addition to increased splash and spray that reduce a driver's visibility when driving in inclement weather.

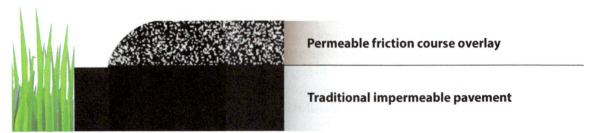

Permeable friction course overlay

Traditional impermeable pavement

Figure 2-23
Graphic depiction of an asphalt permeable friction course overlay over traditional asphalt pavement
Source: © VHB

PFC Applications—Concerns

While PFC consists of smaller aggregates than full-depth porous asphalt, both share the fundamental characteristic of permeability produced by eliminating the fine aggregate from the traditional asphalt mix. This allows precipitation to drain through the asphalt media. Both materials are chosen for their ability to make driving surfaces safer in wet weather conditions, reduce traffic noise and improve discharge water quality. While sharing these similarities, PFC overlays and full-depth porous asphalt also differ in a number of ways.

PFC is applied only as a thin drainage layer of about 25 to 50 mm (1 to 2 in.) in thickness over existing impermeable asphalt roadways with no infiltration into the subsurface. In contrast, the installation of porous asphalt allows stormwater runoff to infiltrate into the ground and recharge groundwater.

Porous asphalt tends to withstand winter conditions better than PFC overlays and poses less of a threat of black ice as water drains downward and away from the surface of the road rather than laterally through the asphalt media. If not properly maintained, both porous asphalt and PFC have the potential to clog because of their permeability. When using PFC, there is potential for water to become trapped in void spaces at the boundary between the PFC and the underlying impermeable roadway. This water can then freeze and become dangerous, forming black ice. In addition, PFC can exhibit temperatures -17° to -16 °C (2° to 4°F) lower than the surface temperature of adjacent conventional asphalt, producing earlier and more frequent frost and ice formation (Liu et al. 2010). Over time, this could promote pavement cracking and raveling.

The result is that more winter maintenance is required for PFC than full depth porous asphalt (Cooley et al. 2009). More salt (or deicing agents) and more frequent applications are required. Spreading of sand to enhance friction and hasten deicing contributes to the clogging of voids, causing a decrease in drainage and noise reduction effectiveness, which negate two of the main PFC advantages (Liu et al. 2010). However, the full depth porous asphalt is not ideal for high speed, high traffic roads, whereas, most PFC applications to date exist on highways.

PFC General Design Criteria—General Specifications

Historically, open-graded asphalt applications posed compromising problems for underlying layers of impermeable asphalt, which tended to undergo stripping at an accelerated rate. The most accepted hypothesis to explain this condition pointed to water that became trapped in the lower portions of the open-graded layer. The ability of this moisture to evaporate was impaired and attacked the surface of the traditional asphalt layer, causing failure (McGhee et al. 2009). This trapped water was and, in some cases, continues to be a cause for concern in cooler climates that experience freeze thaw conditions that lead to black ice on the road surface (Cooley et al. 2009). Additionally, raveling occurred to some original open-graded surface mixes that did not include polymers, which provide flexibility and durability.

Modern applications are addressing the problems of the past by increasing void spaces to allow for better drainage (Cooley et al. 2009) and applying anti-stripping agents like hydrated lime on the existing asphalt or sometimes trackless tack that doubles as a waterproofing membrane and bonding material before installing the porous overlay (McGhee et al. 2009). Also, polymer-modified bonding agents (Roque et al. 2009), asphalt-rubber (Cooley et al. 2009) and liquid-suspending fibers are included in the open-graded mix to reduce raveling (McGhee et al. 2009).

Higher binder contents decrease abrasion loss and thus provide better durability (Liu et al. 2010). Newer PFC mixtures are produced with at least 20% more asphalt binder than original formulas from nearly 40 years ago. The amount of air voids also increased from about 5% to around 20%, making the layer more permeable, which is linked to better safety (i.e., better wet skid resistance) and a longer lifespan (8 to 12 years) (Cooley et al. 2009). However, too much binder content can result in lower air voids content, and thus reducing permeability. A study conducted by Texas A&M (2010) indicated that by increasing asphalt content from 4.5% to 6%, a drop in permeability was observed from a value between 0.5 and 0.55 m³/day (500 and 550 liters/day) to one between 0.4 and 0.425 m³/day (400 to 425 liters/day) (Liu et al. 2010). Compaction also reduces air void content, so particular attention should be paid to the construction specifications during installation of PFC.

Recommended PFC Installation Guidelines

PFC can be installed over existing traditional dense-graded hot mix asphalt (DGHMA) pavement, existing Portland cement concrete pavement or a new HMA binder course. The installation does not fundamentally vary from methods used to apply traditional hot mix asphalt pavements; however, there are certain precautions and specific procedures that should be followed (Cooley et al. 2009). For example, PFC should not be installed if the ambient or pavement surface temperatures are below 10°C (50°F) (McGhee et al. 2009), nor should the mixture be hauled for an extended amount of time to reach the project site. Often, PFC suppliers will limit their haul distance to 80.5 km (50 mi) or 1 hour of traveling time (Cooley et al. 2009).

Before installing a wearing layer of PFC, the existing stable pavement is prepared by removing any particulate matter including dust, mud, or debris and all pavement markings. Cracks wider than 6 mm (0.24 in.) are sealed and irregularities with a depth greater than 25 mm (1 in.) are filled (McGhee et al. 2009). Once general repairs and maintenance are performed, a tack coat should be uniformly applied to the road surface (Cooley et al. 2009). Following these preparatory tasks, the PFC mixture is applied (McGhee, et al. 2009) with a paver that has been calibrated such that the augers turn 85% to 90% of the time. Ideally, the paver should be kept in constant motion with as few stop-and-go interruptions as possible. The mixture is spread over the surface at a thickness of 25 to 50 mm (1 to 2 in.). The spacing for which the lift is tapered at the edges depends upon the type of abutting surface. If a channel runs alongside the road, the PFC can be installed to the edge of the road. When the lift approaches a vegetated strip, at least 100 mm (4 in.) should remain between the PFC and the soft shoulder. In cases where a hard shoulder is adjacent to the road, the PFC layer should be extended to the shoulder by at least 300 to 500 mm (12 to 20 in.) (Cooley et al. 2009).

Similar to full-depth porous asphalt applications, PFC should be compacted with a steel wheel of at least 8 to 10 tons and one to two passes going no faster than 3 mph. Vibrating rollers should not be used in an effort to avoid fracturing the aggregate skeleton of the layer (McGhee et al. 2009) and rubber-tire pneumatic rollers should not be used to avoid pulling up the material (Cooley et al. 2009).

PFC Post-construction Operation and Maintenance

The primary goal of porous asphalt system maintenance is to prevent the pavement surface from becoming clogged by sediment and/or debris. Therefore, the pervious pavement surface should typically be vacuumed twice per year with a commercial street cleaning unit. In areas that receive unusually high amounts of sediment or debris, vacuuming should be done more frequently; in pristine areas, vacuuming may be reduced. Pavement washing systems or compressed air units are not recommended. However, studies have shown good results from vehicles that contain both pressure washing and vacuum equipment to remove accumulated particles in the pavement surface. Some studies speculated that the suction effect on the vehicles tires can clean and protect the pores from clogging at high speeds. However, there is no research that supports or concludes this, so further research is necessary (Liu et al. 2010).

A survey was conducted as part of the TxDOT Project 0-5863 to gather information concerning changes in performance and maintenance noted nationally for PFC and OGFC mixtures. Based on results from this survey, the cleaning of PFC in the United States is not a common practice, which may lead to increasing corrective maintenance and full pavement rehabilitation. Based on the results of the survey, dense-graded hot mix asphalt (DGHMA) is used to repair delaminated areas and potholes in PFC by all states in the United States that reported using PFC in 2000 (Liu et al. 2010). If the area to be repaired is small and the flow around the patch can been ensured, DGHMA is recommended for patching. Otherwise, the area should be repaired using the PFC mixture, as patching and crack filling with DGHMA may generate drainage problems since flow inside the mixture is diminished (Estakhri et al. 2008). General recommendations and actual practices for the rehabilitation of PFC in the United States includes milling and replacing the existing PFC with new PFC or conventional asphalt. The direct placement of DGHMA over PFC is not recommended because life of the new layer can be diminished by water accumulation inside the PFC (Liu et al. 2010).

Although the structural life of the PFC is expected to be between 8 to 12 years, as previously indicated, the functional life of PFC is expected to be between 5 and 8 years. Functionality is affected by air voids content reductions during service as a consequence of clogging. Therefore, in the absence of cleaning activities, the initial permeability and noise reduction capacity are expected to decrease such that, at the end of the functional life, PFC behaves much like dense-graded mix (Estakhri et al. 2008).

PFC Cost Information

PFC is an application of porous asphalt. Porous asphalt with additives is generally 20% to 50% higher in cost than conventional asphalt on a unit area basis (for the pavement itself). See **Section 2.6 Cost Information** for additional information on porous asphalt costs.

PFC Benefits

Pollutant Removal

Water quality can be improved in areas where PFC is installed. Instead of rain falling on an impermeable surface and discharging as sheet flow, carrying with it pollutants from the roadway surface, rain has to first travel through a network of voids in the PFC, where particle matter can be trapped in the pore space of the pavement. In effect, the permeable layer works as a filter system for stormwater runoff. Studies in highway applications have shown that PFC can provide stormwater runoff mitigation as well by significantly reducing the amount of pollutants discharged from paved areas.

In a study conducted between February 2004 and December 2007 (Barrett 2008), the quality of stormwater that passed through a permeable friction course overlay on a highway in Austin, TX was found to be significantly improved from that of stormwater runoff from impermeable asphalt pavements. Runoff samples were taken during individual storm events occurring before (five samples) and after (thirty-six samples) installation of PFC on two lanes of a four-lane highway over a 40-month period.

The samples were brought to a lab, where the concentrations of a broad spectrum of pollutants were measured. After comparing the concentrations from samples of runoff gathered from the impermeable asphalt against those from the new PFC overlay, the results showed a significant decrease in concentrations of particles and particle-associated pollutants including total suspended solids (TSS) (93%), total Cu (52%), total Pb (88%) and total Zn (81%). The concentrations of nitrate/nitrite, dissolved copper and zinc, and total and dissolved phosphorus did not exhibit a significant difference between the two road surface, which suggests PFC has little to no effect upon concentrations of dissolved constituents in stormwater runoff (Barrett 2008). A second study was conducted on the same highway 1.5 km (1 mile)away and showcased a side-by-side application of PFC and conventional pavement (thirteen samples each of PFC and standard asphalt). The study showed similar results.

Pagotto et al. (2000) also studied the effects of PFC on the quality of highway runoff at a section of a highway in France. The site was originally conventional pavement and was replaced with a porous asphalt surface. Data was collected over for each pavement type and the results showed a significant reduction in total suspended solids (TSS) (81% reduction). Total metals were all reduced: Pb by 78%, Cd by 69%, Zn by 66%, and Cu by 35%. Some dissolved metals (Zn and Cd) were also reduced by about 60%. Hydrocarbons were reduced by 92% (Barrett 2008).

Barrett proposed two potential reasons for the improved stormwater quality associated with attributes of the PFC. Due to the permeable nature of the overlay, splash and spray are reduced during rain events, which may decrease the chances of pollutants from the underside of vehicles being displaced and carried with rainwater. Also, pollutants that do get carried with the stormwater runoff are filtered through the permeable layer before reaching the sides of the road. This provides a chance for particle pollutants to become trapped in the void spaces of the layer. As particles and particle-associated pollutants accumulate within the pore structure of the PFC, the layer may become clogged. This increases the potential for runoff to travel on the surface of the pavement, resulting in concentrations that may not be significantly reduced from conventional pavement (Barrett 2008). To keep the stormwater quality benefits of the PFC intact, maintenance should be performed to remove accumulated material in the PFC.

Noise Pollution Reduction

The void spaces at the surface of PFC also reduce noise generated from tire-pavement contact. Sound waves are intercepted by the aggregate's pores in the PFC, thereby retarding the ability of the sound waves to travel long distances. According to the National Center for Asphalt Technology at Auburn University, the porosity and rubber or polymer modifiers found in PFC overlays significantly reduce the noise generated from tire-pavement contact (Rasmussen et al. 2007). Some researchers found that the thicker the PFC layer, the lower the noise levels at the tire/pavement surface, while other studies reported that noise levels increase with an increase in surface layer thickness (Liu et al. 2010). Further research on the properties of PFC as it relates to noise reduction is merited to fully understand and harness the benefits of PFC on noise reduction.

Based on the results of a study done on noise testing of the PFC Pavements in Texas (Trevino-Frias and Dossey 2007), results indicate that overall sound levels of PFCs averaged 98.1dBA, while the overall sound level for other pavements is 101 dBA. This represents a potential reduction of almost 3 dBA, which corresponds to a halving of traffic volume, assuming that most of the roadside noise is generated by pavement/tire contact. Other studies referenced in this report found similar results with 8 to 10 dBA reduction (Noise Intensity Testing in Europe [NITE] study, funded by Caltrans) and 4.5 dBA reduction (University of California [UC] David study) for PFC pavements as compared with other dense-graded asphalt mixes.

2.9 Resources for Specifications

Porous Asphalt

Specifications have become increasingly available in recent years although none are likely to be applicable to the entire US without some modifications to local conditions. A partial list of specifications/guidance includes:

- ASTM WK15789—New Practice for Construction of Porous Asphalt Pavements with Stone Reservoirs (a committee is currently working on this standard)
- Minnesota Asphalt Pavement Association, "DRAFT Guidance Specification for Porous or Dense-Graded Hot-Mix Asphalt Pavement Structures for Storm Water Management," <www.asphaltisbest.com>
- National Asphalt Pavement Association (NAPA): Information Series (IS)-131 Porous Asphalt Pavements (2008) <www.hotmix.org>
- Iowa Stormwater Management Manual, Version 3, Section 2J-3, Porous Asphalt, October 28, 2009.
- University of New Hampshire Stormwater Center (UNHSC) <http://www.unh.edu/erg/cstev/pubs_specs_info/unhsc_pa_spec_10_09.pdf>.
- Some state Department of Transportation (DOT) specifications may also be partially applicable for porous asphalt systems although they should be reviewed carefully before being used. Notable examples include:
  - Oregon DOT F-Mix
  - Iowa Stormwater Management Manual, Version 3, Section 25-3, Porous Asphalt, October 28, 2009.

Permeable Friction Course Overlays

Specifications have become increasingly available in recent years although none are likely to be applicable to the entire US without some modifications for local conditions. A partial list of specifications/guidance includes:

- ASTM D7064 Standard Practice for Open-Graded Friction Course (OGFC) Mix Design
- University of New Hampshire Stormwater Center (UNHSC) <http://www.unh.edu/unhsc/>
- National Asphalt Pavement Association (NAPA): Information Series (IS)-131 Porous Asphalt Pavements (2008) <www.hotmix.org>
- NAPA IS-115 Open-Graded Friction Courses (2002) <www.hotmix.org>
- ASTM WK-15789 (a committee is currently working on a specification for porous asphalt)
- See "Standard Practice for Materials, Design and Construction of Permeable Friction Courses (PFC)" from AASHTO (still in Draft)
- Some state Department of Transportation (DOT) specifications may also be partially applicable for porous asphalt systems although they should be reviewed carefully before being used. Notable examples include:
  - Oregon DOT F-Mix
  - Georgia DOT Open-Graded Friction Course
  - Texas DOT—Item 342 Permeable Friction Course (PFC)

2.10 General Resources

Ferguson, B. (2005). "Porous Pavements." CRC Press, Boca Raton, FL.

National Asphalt Pavement Association (NAPA). Hansen, K. (2008). "Porous Asphalt Pavements for Stormwater Management; Design, Construction and Maintenance Guide." National Asphalt Pavement Association.

University of New Hampshire Stormwater Center (UNHSC) (2010). "University of New Hampshire Stormwater Center 2009 Annual Report." Durham, NH. <http://www.unh.edu/unhsc/>.

SEMCOG (2008). "Chapter 7: Structural Best Management Practices." *Low Impact Development Manual for Michigan.*" <www.semcog.org/lowimpactdevelopmentreference.aspx>.

2.11 References

Alvarez, A. E., and Martin, A. E. (2009). "Permeable Friction Course Mixtures are Different." *Proc., Rocky Mountain Asphalt Conference & Equipment Show*, Colorado State University, Fort Collins, CO.

Backström, M. (1999). "Porous Pavement in a Cold Climate." Licentiate Thesis. Lulea, Sweden: Lulea University of Technology.

Barrett, M. E. (2006). "Stormwater Quality Benefits of a Permeable Friction Course." *World Environmental and Water Resources Congress 2006: Examining the Confluence of Environmental and Water Concerns*, ASCE, Reston, VA.

Barrett, M. E., and Shaw, C. B. (2007). "Stormwater Quality Benefits of a Porous Asphalt Overlay." *Transportation Research Record: Journal of the Transportation Research Board.*

Barrett, M. E. (2008). "Effects of Permeable Friction Course (PFC) on Highway Runoff." *Proc., 11th International Conference on Urban Drainage*, Edinburgh, Scotland, UK.

Bean, E., Hunt, W., and Bidelspach, D. (2007). "Field Survey of Permeable Pavement Surface Infiltration Rates." *Journal of Irrigation and Drainage Engineering.* 133(3), 249–255.

Cooley, L. A., Brumfield, J. W., Mallick, R. B., Mogawer, W. S., Partl, M., Poulikakos, L., et al. (2009). "Construction and Maintenance Practices for Permeable Friction Courses." *National Cooperative Highway Research Program Report 640*, Transportation Research Board, Washington, D.C.

Estakhri, C. K., Alvarez, A. E., Martin, A. E. (2008). "Guidelines and Construction and Maintenance of Porous Friction Courses in Texas." Texas Transportation Institute, Texas A&M University, Report 0-5262-1.

Ferguson, B. (2005). "Porous Pavements." CRC Press, Boca Raton, FL.

Gunderson, J. (2010). "Boulder Hills LID Economic Case Study." *Forging the Link Between Research-Based Institutions, Watershed Assistance Groups, and Municipal Land Use Decisions*, UNH Stormwater Center, Durham, NH.

Liu, K., et. al. (2010). "Synthesis of Current Research on Permeable Friction Courses: Performance, Design, Construction, and Maintenance." Texas Transportation Institute, Texas A&M University, Report 0-5836-1.

McGhee, K., Clark, T., and Hemp, C. (2009). "Research Report: A Functionally Optimized Hot-Mix Asphalt Wearing Course: Part 1: Preliminary Results." Virginia Transportation Research Council, Charlottesville, VA., Final Report VTRC 09-R20 <http://www.virginiadot.org/vtrc/main/online_reports/pdf/09-r20.pdf>.

Muench, S. T., and Mahoney, J. P. (2002). "Washington Asphalt Pavement Association Asphalt Pavement Guide: Pavement Types." Washington Asphalt Pavement Association, Inc., <http://www.asphaltwa.com/wapa_web/index.htm> (Dec. 16, 2009).

National Asphalt Pavement Association (NAPA). Hansen, K. (2008). "Porous Asphalt Pavements for Stormwater Management; Design, Construction and Maintenance Guide." National Asphalt Pavement Association.

Pagotto, C., Legret, M., and Le Cloirec, P. (2000). "Comparison of the Hydraulic Behaviour and the Quality of Highway Runoff Water According to the Type of Pavement." *Water Resources*, 34(18), 4446-4454.

Rasmussen, R. O., Bernhard, R. J., Sandberg, U., and Mun, E.P. (2007). "The Little Book of Quieter Pavements." US Department of Transporation Federal Highway Administration, FHWA-IF-08-004, Washington, D.C.

Roque, R., Koh, C., Chen, Y., Sun, X., and Lopp, G. (2009). "Introduction of Fracture Resistance to the Design and Evaluation of Open-graded Friction Courses in Florida." Florida Department of Transportation and University of Florida, Tallahassee, FL.

Roseen, R. M., Ballestero, T. P., Houle, J. J., Briggs, J. F., and Houle, J. P. (2012). "Water Quality and Hydrologic Performance of a Porous Asphalt Pavement as a Stormwater Treatment Strategy in a Cold Climate." *ASCE Journal of Environmental Engineering.* 138(1), 81–89, <doi: http://dx.doi.org/10.1061/(ASCE)EE.1943-7870.0000459>.

Roseen, R. M., and Ballestero, T. P. (2008). "Porous Asphalt Pavements for Stormwater Management in Cold Climates." *HMAT*, National Asphalt Pavement Association, 13(3).

Texas Department of Transportation (2004). "TxDOT Specifications." *Standard Specifications for Construction and Maintenance of Highways, Streets, and Bridges*, Item 342, 312–329,<http://www.dot.state.tx.us/business/specifications.htm> (Mar. 20, 2005).

Thelen, E., and Howe, L. F. (1978). *Porous Pavement,* Franklin Institute Press, Philadelphia, PA.

Transportation Research Board (TRB) (1992). "*Porous Pavement for Control of Highway Runoff in Arizona: Performance to Date.*" Transportation Research Board, The National Academies, Washington, D.C., 45–54, <http://trid.trb.org/view.aspx?id=370836>.

Trevino-Frias, M., and Dossey, T. (2007). "Preliminary Findings from Noise Testing on PFC Pavements in Texas." *Noise Level Adjustments for Highway Pavements in TxDOT,* Center for Transportation Research at The University of Texas at Austin, Report 0-5185-2, <http://www.utexas.edu/research/ctr/pdf_reports/0_5185_2.pdf> (Oct. 24, 2013).

University of New Hampshire Stormwater Center (UNHSC) (2009). *University of New Hampshire Stormwater Center 2009 Annual Report*, Durham, NH. <http://www.unh.edu/unhsc/>.

University of New Hampshire Stormwater Center (UNHSC) (2009). "Fact Sheet: Porous Asphalt Pavement for Stormwater Management." <http://www.unh.edu/unhsc/>. (December 16, 2009).

United States Environmental Protection Agency (US EPA) (2012). "Hurd Field Porous Pavement Education Project." United States Environmental Protection Agency, <http://www.epa.gov/mysticriver/porouspavementproject.html>.

Pervious Concrete Fact Sheet

DESCRIPTION

Pervious concrete consists of a hydraulic cementitious binding system combined with an open-graded aggregate to produce a rigid, durable pavement. This pavement typically has 15% to 25% interconnected void space that allows rapid infiltration of stormwater to the underlying soil and/or aggregate storage layer.

Pervious concrete in San Diego County, CA
Source: CH2M HILL

Pervious concrete in Seattle, WA
Source: Amy Rowe

APPLICATIONS

POTENTIAL APPLICATION		NOTES
Overflow Parking	Yes	
Primary Parking Areas (most heavily used)	Yes	
Sidewalks/Pathways	Yes	
Drive/Aisles	Yes	
Roads/Highways	Limited	Not yet recommended for highway use in the United States
Access Drives/Ring Roads	Yes	
Loading Areas	Limited	Tendency to ravel and deteriorate under high turning loads
Frequent Truck Traffic	Limited	Tendency to ravel and deteriorate under high turning loads

Pervious Concrete Fact Sheet continued

BENEFITS

Primary benefits include stormwater volume reduction, impervious cover reduction, peak flow rate reduction, groundwater recharge, pollutant removal, heat island mitigation, noise reduction, and skid resistance. Stormwater benefits are directly associated with the ability of pervious concrete to drain rainfall to the base/subbase and to infiltrate runoff into underlying soils or convey it to stormwater systems via drains. Draining stormwater through the pavement system prevents water from accumulating on the surface, as it does on traditional pavements, providing additional skid and slip resistance. The filtration through the pavement surface and subbase can also reduce runoff peaks and provide water quality treatment.

STORMWATER QUANTITY FUNCTIONS		NOTES
Volume	High	Volume control up to 100% with total infiltration; may mitigate for greater area with additional aggregate storage layer. Installation permeability 750–5,000 cm/hr (300–2,000 in./hr).
Groundwater Recharge	High	Recharge high if soils allow
Peak Rate	Medium/High	Peak discharge rate control increased with increased storage and/or high permeability in existing soils
STORMWATER QUALITY FUNCTIONS		**NOTES**
Total Suspended Solids	High	Pre-treatment suggested for run-on to pervious pavement from contributing area
Total Nitrogen/Total Phosphorus	Moderate	Pre-treatment suggested for run-on to pervious pavement from contributing area
Metals	High	
Temperature	High	Typically maintains and does not increase ambient runoff temperature, except where radiant heating increases thermal mass

SITE CONSTRAINTS/CONCERNS

- For infiltration and microbial activity, a minimum 60 cm (2 ft) of separation from bedrock and seasonal high groundwater elevation is typical.

- Low soil infiltration rates [below 12 mm/hr (0.5 in./hr)] may require additional design considerations and engineering (e.g., overflow systems, storage bed thickness, and underdrains).

- Local geology (especially in karst [sinkhole-prone] settings)

- Separation from wells, septic systems, subsurface structures (e.g., basements), etc.

- Land use considerations with concern related to potential for groundwater contamination

- Cold climate and deicing chemical effects on performance and maintenance including sanding with fine particles and road salt

- Subgrade disturbance should be minimized during construction to reduce compaction and promote infiltration.

- Hot and cold weather placement may have specific restrictions.

Pervious Concrete Fact Sheet continued

RECOMMENDED KEY DESIGN CRITERIA

- Voids in pervious concrete surface layer typically 15% to 25%
- Voids for the aggregate storage base/subbase typically 36% to 42%
- Crushed aggregate should be used for the reservoir base/subbase.
- Load-bearing and infiltration capacities of soil subgrade
- Load-bearing and infiltration capacity of the pervious concrete
- Storage capacity of the stone base/subbase. A greater thickness may be required for the aggregate base in cold climates.
- Conveyance of larger storm event flows to stormwater system
- Underdrains and/or liners may be required for non-ideal soils
- Drain at higher elevations may be required for additional detention
- Protection from high sediment loads associated with run-on (i.e. pre-treatment) and contributing area run-on ratio
- Requires a minimum curing time of 7 days (14 days for truck traffic)
- Should not be installed when ambient temperature is 4°C (40°F) or lower

PERFORMANCE

MEDIUM—Pervious concrete can be designed to handle heavy loading; however, standard pavement surfaces should be used in areas of heavy truck traffic and/or high turning frequency due to the tendency of permeable pavements to ravel and deteriorate under high turning loads.

OPERATION & MAINTENANCE

MEDIUM—Vacuum twice per year with a commercial vacuum sweeper and more frequently for areas that receive unusually high amounts of sediment or debris. A practical, but rigorous, erosion and sedimentation control plan should be developed to protect from run-on sediment or debris that may cause clogging. Adjacent vegetated areas should be well maintained. Bare spots or eroded areas should be immediately replanted and stabilized.

Snow plowing is permissible for pervious concrete; however, the use of deicing materials and sand should be limited. Typically, less deicer and sand are required due to the infiltration of melt-water through the surface voids and reduced chance of refreezing. Deicing materials can damage the cement in the pervious concrete mixture and result in disintegration or spalling of the surface, particularly if used over a newer application. Deicers should never be applied within the first year of installation. Deicers can be used in limited amounts on older applications. Certain weather conditions, in which the ground becomes very cold, can allow build up of snow and ice on the surface. In those instances, significantly more deicing material may be required because the applied brine will infiltrate and not pond on the surface. The use of sand will reduce the surface infiltration rate and require vacuuming.

Pervious Concrete Fact Sheet continued

COSTS

MEDIUM/HIGH—Costs vary significantly depending upon the particular application. Pavement, storm drainage, soils and slope considerations all impact the installation costs of the pavement. Generally, the cost for pervious concrete is higher than the cost for traditional concrete, but results in savings when the storm drainage requirements are included. Overall, the maintenance costs associated with pervious concrete are lower than those costs for impervious pavements since they do not include catch basins, pipes, ditches, vaults, or ponds components.

REFERENCES

Kevern, J., and Zufelt, J. (2013). "Introduction to ASCE Monograph-Permeable Pavements in Cold Climates." *ISCORD 2013: Planning for Sustainable Cold Regions*, Zufeit, J. E. ed., Amercian Society of Civil Engineers, 482–494, doi: 10.1061/9780784412978.047, <http://dx.doi.org/10.1061/9780784412978.047>.

United States Environmental Protection Agency (US EPA) (2009). "Pervious Concrete Pavement Fact Sheet." <http://water.epa.gov/polwaste/npdes/swbmp/Pervious-Concrete-Pavement.cfm>.

3 Pervious Concrete

3.1 System Description

Pervious concrete consists of a hydraulic cementitious binding system combined with an open-graded aggregate to produce a rigid pavement with typically 15% to 25% interconnected void space. The ultimate goal is to create a durable wearing pavement surface that allows rapid infiltration of stormwater to the underlying soil or open-graded base aggregate layer. The clean aggregate layer typically has 40% compacted voids. Layer depth is determined based on hydrologic design, vehicle loading, and frost depth considerations. A depth of 30 cm (12 in.) minimum in freeze-thaw climates is typical. **Figure 3-1** shows a typical pervious concrete system section. Pervious concrete blocks have been used since the mid-1800s in England for residential construction. Pervious concrete pavements were first used in the Pacific during World War II to help recharge aquifers on allied-occupied islands and for tank facilities in England.

Pervious concrete strength ranges from low for applications not requiring much strength (such as erosion control) to high for specially designed heavy traffic sections. Typical compressive strengths for pervious concrete pavements with around 20% voids are appropriate for parking areas that experience moderate truck traffic. Special admixtures and increased design thicknesses are typically required for heavier applications.

National guidelines on materials, design, and construction have emerged from Committee 522 on Pervious Concrete of the American Concrete Institute (ACI) and material testing standards from the American Society for Testing and Materials (ASTM) Subcommittee C09.49 on Pervious Concrete. The National Ready

Pervious concrete
Typically 20% voids; 10–20 cm (4–8 in.) thick

Open-graded aggregate base
Typically 40% voids; Depth varies

Non-woven geotextile fabric (may be required)

Uncompacted soil subgrade

Figure 3-1
Typical pervious concrete pavement system section
Source: © VHB

Mixed Concrete Association established a Pervious Concrete Contractor Training Course and Certification. These committees and programs have sponsored research and development in many areas of pervious concrete. They have resulted in technical guidance, material and construction specifications, and construction training to facilitate successful projects. Current standards include fresh unit weights and voids (ASTM C1688), field infiltration (ASTM C1701), raveling potential (ASTM C1747) and hardened unit weight and void content (ASTM C1754).

Pervious concrete pavements are most often utilized for stormwater management in vehicle parking areas. However, pervious concrete has been utilized in sidewalks, trails, pathways, residential driveways, residential or low volume neighborhood streets, curbs, gutters, overlays, slope stabilization, erosion protection and repair for bridge abutments, pool deck areas, sound mitigating walls, ditch lining, highway shoulders, floors in greenhouses and horse barns, and drainable layers underneath traditional pavements as well as artificial turf. Pervious concrete has been successfully applied in desert climates in Arizona and Nevada to roadways in Minnesota.

Pervious concrete can be produced using a wide variety of aggregate sizes and shapes, although 4.75 to 9.5 mm (0.25 to 0.5 in.) coarse aggregate has shown to produce a desirable surface texture, and 5.75 mm (0.23 in.) is the most commonly used size. **Figure 3-2** illustrates the texture of pervious concrete pavement. The permeability of newly installed pervious concrete typically ranges from around 500 cm/hr (200 in./hr) to over 7,600 cm/hr (3,000 in./hr). **Figure 3-3** illustrates rapid surface infiltration of water from a hose.

Figure 3-2
Typical pervious concrete surface, Kansas City, MO
Source: University of Missouri-Kansas City

Figure 3-3
Demonstration of rapid infiltration, North Liberty, IA
Source: John Kevern

3.2   Applications—Site Constraints/Concerns

Pervious concrete is well suited for low to moderate traffic parking areas, sidewalks, pathways, recreational areas (e.g., basketball courts), driveways, and low-speed/low-volume roadways (see **Figures 3-4 through 3-13**). Higher volume roads and highways have also been constructed of pervious concrete in Europe, Japan and Australia with success. In the United States, there are more limited applications of pervious concrete on heavier traffic volume areas. Although, research is helping to move toward more wide-spread utilization. A pervious concrete overlay in Minnesota is one of the quietest concrete

Figure 3-4
Pervious concrete adjacent to traditional concrete during a rainfall, Ames, IA
Source: John Kevern

Figure 3-5
Pervious concrete sidewalk at the Johnson County Community College, Overland Park, KS
Source: Concrete Promotional Group, Inc.

Figure 3-6
Pervious concrete at Kansas City Zoo, Kansas City, MO
Source: Concrete Promotional Group, Inc.

Figure 3-7
Pervious concrete residential driveway, Kansas City, MO
Source: John Kevern

Figure 3-8
Pervious concrete at the Regional Athletic Complex, Lacey WA
Source: Puget Sound Concrete Specification Council

Figure 3-9
Pervious concrete overlay at a MnROAD Facility, Albertville, MN
Source: John Kevern

Figure 3-10
Pervious concrete as a drainage layer underneath playground, Fort Riley, KS
Source: Concrete Promotional Group, Inc.

Figure 3-11
Completed playground with artificial grass over pervious concrete, Fort Riley, KS
Source: Concrete Promotional Group, Inc.

Figure 3-12
Pervious concrete on an industrial site, Shelter Systems, MD
Source: National Ready Mix Concrete Association

Figure 3-13
Pervious concrete at a Fire Station, SeaTac, WA
Source: Puget Sound Concrete Specification Council

pavements in the United States (Schaefer 2010) and the Missouri DOT is constructing a roadway section near St. Louis with a photocatalytic, self-cleaning pervious concrete shoulder for stormwater management, smog reduction and urban heat island mitigation.

Uses of pervious concrete can be grouped into the following categories:

Pavement:

- Parking areas
- Minimal traffic applications (e.g., sidewalks, paths, patios, etc.)
- Highway shoulders
- Full-depth roadways
- High volume wearing course overlays

Non-pavement:

- Curb and gutters
- Hydraulic structures
- Sound mitigating structure
- Erosion protection

Climate Considerations

- **Figure 3-14** shows pervious concrete placement adjacent to traditional concrete in a cold climate application. The voids in the pervious concrete surface allow for rapid infiltration of surface water, which decreases the risk of refreezing and icing. The voids also allow for faster melting of ice and snow on the surface. As a result, the need for deicing materials and sand application is decreased.

- In general, the use of deicing materials and sand is not recommended on pervious concrete. If necessary, the use of these materials should be limited. Deicing materials can damage the cement in the pervious concrete mixture and result in disintegration or spalling of the surface. Deicers should never be applied within the first year of installation. During this time, curing of

Figure 3-14
Standard concrete adjacent to pervious concrete in cold climate, Iowa State University, Ames, IA
Source: John Kevern

the surface is still occurring and can be significantly damaged by exposure to deicers. If sand is used, vacuum cleaning of the surface should be performed at the end of the winter season.

- Many specifications require that the contractor not place pervious concrete for pavement when the ambient temperature is 4°C (40°F) or lower. The ready mixed concrete supplier may be required to use heated mixing water for winter, unless a cold weather plan is submitted and permitted in writing by the specifier. Hot water accelerates setting and reduces working times, which is not desirable.

- Because of the chemical admixtures and the open voids in pervious concrete, setting time is delayed and large amounts of heat typical to conventional concrete curing are not generated during hydration. Therefore, hot water is not beneficial and blankets should not be used to capture curing heat. Blankets should only be used to protect from freezing should the temperature drop during the curing phase.

- Restrictions may be required for hot weather placements due to high evaporation climate and low water/cement ratio, which leads to rapid drying. For further information, refer to ACI 305 *Guide to Hot Weather Concreting*.

3.3   General Design Criteria—General Specifications

General Design Considerations

If intended to achieve significant stormwater infiltration, pervious concrete systems should follow the general design guidelines in **Chapter 1** and the following:

Structural Design Considerations

- Pervious concrete is a rigid pavement and should be designed accordingly. In most cases, the depth of the aggregate storage layer selected during the hydrologic design will be greater than the base required for traffic loading, but both should be considered and the thicker base used.

- Strength and permeability are inversely related for permeable pavements. At a particular void content, the strength of pervious concrete can be increased either by using a stronger aggregate with more fractured faces for better locking, improving the aggregate gradation or improving the strength of the cementitious paste.

- Tests using a falling weight deflectometer and analysis have shown that for one installation, a 150 mm (6 in.) thick pervious concrete layer with 450 mm (18 in.) of crushed limestone aggregate base overlaying saturated glacial till is a stronger pavement section than 150 mm (6 in.) of traditional concrete placed directly on compacted glacial till (Suleiman 2011).

- The pervious concrete industry is developing ASTM test methods for characterizing compressive or flexural strengths of pervious concrete as well as other properties. These tests are needed to model fatigue of rigid pavement under loads. As an interim step, fatigue equations published by the American Concrete Pavement Association (Rodden et al. 2010a) assume such inputs to be comparable in nature—but not magnitude—to those used for conventional rigid concrete pavements. The ACPA design method should be consulted for further information (Rodden et al. 2010b). Other references on structural design include Delatte (2007) and Goede (2011).

- It is not recommended to have filter fabrics directly beneath the pervious concrete surface material as it can collect fines and reduce permeability. For structural support underneath the aggregate base, geotextiles may be used as directed by the engineer.

- General guidelines for pervious concrete surface thickness (as a function of traffic and soil type) are published by the National Ready Mix Concrete Association and the Portland Cement Association (Leming 2007).

Hydrologic Design

The hydrologic design of pervious concrete is similar to that of other permeable pavements. See **Chapter 1 and Chapter 9**. Additional references include Leming (2007) and the Portland Cement Association (PCA 2007).

Pervious Concrete Materials

The objective of pervious concrete mixture proportioning is to produce a permeable, smooth, and durable pavement surface, typically achieved at around 20% voids. The voids are created by starting with a narrowly-graded coarse aggregate and balancing the cementitious paste volume in it for durability. The objective is to create a mix that maintains permeability and strength with sufficient paste viscosity to allow the mixture to be workable without paste drain down. The correct paste viscosity is achieved using a lower water-to-cement ratio (0.27 to 0.34) than conventional concrete with chemical admixtures.

To maintain workability, typically high range water-reducing admixtures are combined with hydration stabilizing admixtures. Additional admixtures such as viscosity modifying or latex-based products can be included for further modification of the paste and mortar properties. Air entraining admixtures and latex additives have been shown to increase the freeze-thaw durability of pervious concrete. Field testing air content as a result of air entraining admixtures is not practical. Air entrainer dosages are similar to other low-slump concrete such as curb and gutter mixtures. A small fraction of fine aggregate (sand) has shown to increase strength and freeze-thaw durability. Many pervious concretes contain 5% to 7% sand by weight of the total aggregate, with some containing up to 10% in addition to that contained in the coarse aggregate gradation.

Polypropylene or cellulose fibers are commonly used in pervious concrete for improved durability and workability (Kevern 2013). Fibers tend to link up void spaces and produce higher permeability than non-fiber mixtures with the same void content. Fibers have been successful in remediating marginal mixtures. For wet mixtures with the potential to become over-compacted, the "bird's nest" effect of the fibers helps maintain the open void structure. For dry mixtures with the potential for excessive raveling, fibers help maintain the surface integrity (Kevern 2013). Additional resources on pervious concrete mixture proportioning include Schaefer (2006) and Kevern (2011).

Reservoir Course

The reservoir layer beneath the pervious concrete surface layer should be composed of clean, open-graded aggregate, typically AASHTO No. 57, which has an average size of 19 mm (0.75 in.) (<2% passing ASTM No. 200 sieve). Typically, void space of this layer ranges from 36% to 42%, which is determined (per ASTM C29) on a dry-rodded basis. The thickness of this layer varies, and is determined based on stormwater management and structural requirements. Low traffic applications, such as sidewalks and pathways, may not require any reservoir course layer. Heavier traffic applications typically require reservoir depths ranging from 200 to 300 mm (8 to 12 in.) Thicker aggregate base layers may be required in very cold climates to prevent from freezing in the pavement surface void spaces. A minimum of 300 mm (12 in.) to 450 mm (18 in.) is recommended in cold climates.

3.4   Recommended Installation Guidelines

For detailed specifications, practitioners should refer to ACI 522.1 *Specifications for Pervious Concrete Pavement*. This guide specification covers materials, preparation, forming, placing, finishing, jointing, coring, and quality control of pervious concrete. Provisions for testing, evaluation and acceptance of pervious concrete are included in this specification. Components of this installation specification are outlined on the following pages.

Recommended Outline for Installation: Pervious Concrete

A. Subgrade and Reservoir Course

1. **Subgrade**

   - Comply with specified density requirements for compacted soil subgrades and do not over compact.

   - Existing soil subgrades and cuts to establish subgrade elevations and slopes shall not be subject to repeated construction equipment traffic prior to reservoir layer placement. Scarify to improve infiltration rates (**Figure 3-15**).

   - When soil fill is needed to establish proper subgrade level, some compaction may be necessary. Compact to a minimum of 92% of standard Proctor density (NRMCA 2010).

   - Fill and re-grade any areas damaged by erosion, ponding, or traffic compaction before the placing (optional) geotextile and the aggregate reservoir layer.

2. **Installation of (optional) geotextile and aggregate reservoir layer**

   - Upon completion of subgrade preparation, notify the architect/engineer to inspect before the contractor proceeds with the reservoir layer installation.

   - Place the (optional) geotextile and the reservoir layer aggregate immediately after approval of subgrade preparation. Remove any accumulation of debris or sediment after approval of subgrade prior to installation of geotextile at the contractor's expense (**Figure 3-16**).

   - Place (optional) geotextile in accordance with manufacturer overlap recommendations. Adjacent strips overlap a minimum of 40 cm (16 in.). Geotextile shall be placed on the horizontal soil subgrade and on the vertical sides of the pavement base and surface. The contractor shall secure fabric at least 60 cm (2 ft) outside of bed and prevent any runoff or sediment from entering the storage bed.

   - Install coarse aggregate in minimum 150 mm (6 in.) thick lifts. Compact each lift to the specified density. Install aggregate to grades as indicated on the drawings (**Figure 3-16**).

   - Place compost socks, hay bales, silt fences, or geotextile adjacent to storage beds to prevent sediment from washing into beds during site development. As the site is fully stabilized, excess geotextile along the pavement perimeter can be cut back to the edge of the pavement.

Figure 3-15
Excavation of base for aggregate storage layer prior to scarification and placement of geotextile fabric, Ames, IA
Source: John Kevern

Figure 3-16
Aggregate base after installation of geotextile fabric, Ames, IA
Source: John Kevern

B. Pervious Concrete Layer

Note: Off-site test placements for the contractor and supplier are highly recommended for quality purposes, especially for first time or non-certified installers. Typically, a 3 x 6 m (10 x 30 ft) test strip is adequate. When possible, bidders should be prequalified based on training and pervious concrete installation experience, including oversight by a craftsman-level contractor.

1. **Pavement thickness**: Pavement shall be placed to the depth as indicated on the drawings.

2. **Formwork**: Form materials shall be wood, steel, or other and shall be the full depth of the pavement. Forms shall be of sufficient strength and stability to support mechanical equipment without deformation during installation and shall conform to plan profiles following spreading, strike-off, and compaction operations (**Figure 3-17**).

3. **Mixing and hauling**

   - Production: Manufacture and deliver pervious concrete accordance with ASTM C94.

   - Mixing: Produce mixtures in central mixers or in transit (truck) mixers. Mix concrete for a minimum time specified according to ASTM C94.

   - Transportation: The pervious concrete mixture transported and the discharge of individual loads at the site shall be completed within one (1) hour of the introduction of mix water to the cement.

     Note: Delivery times may be extended to 90 minutes when dosages of hydration stabilizer are increased to maintain the concrete or at the discretion of the specifier. Typical admixtures used in pervious concrete function effectively for up to 30 minutes. They start to lose effectiveness between 30 and 60 minutes mixing time.

   - Discharge and Adjustments: Visually inspect each truckload for consistency of concrete mixture. Permit water addition to adjust the consistency at the point of discharge. Subsequent water additions for workability should be adjusted accordingly at the batching facility. Deposit concrete as close to its final position as practical, and such that discharged concrete is incorporated into previously placed plastic concrete (**Figure 3-18**).

Figure 3-17
Forms and tools in place before concrete arrives, Seattle, WA
Source: Puget Sound Concrete Specification Council

Figure 3-18
Pervious concrete discharging into the belt-placer (telebelt) hopper, Lacey, WA
Source: Puget Sound Concrete Specification Council

4. **Placing and finishing**

- The underlying aggregate base layer shall be in a moist condition at time of placement. The contractor shall spray the base layer with water immediately prior to placing pervious concrete.

- Deposit concrete shall be deposited into the forms by mixer truck chute, conveyor, or buggy.

Note: Pervious concrete is **not** pumpable.

- Hand place and with a roller screed properly, weighted with water or sand inside the roller. Failure to weigh down the roller screed will result in poor compaction, poor durability, raveling, and a poor riding surface.

- Other construction methods are available. For example, the contractor may place the pervious concrete with an alternate method approved by the architect/engineer in writing. The contractor must show proper information to substantiate an ability to successfully place pervious concrete with the alternate method.

- Use cross rollers behind the roller screed to aid the rolling out of minor surface imperfections for final compaction.

- **Jointing**

  - Joints shall be constructed utilizing a rolled joint former, also known as a "pizza cutter." Perform jointing immediately after roller compaction. Joints shall be marked where they will be placed before paving starts.

  - Use isolation joints when abutting fixed vertical structures such as manholes, light poles, sign poles, etc. Joints may also be raw cut. Sawing may be performed as soon as the pavement is strong enough to resist raveling and before random cracking occurs.

- **Curing**

  - Begin curing procedures 3 to 5 minutes behind the roller screed.

  - Cover the pavement surface with heavy duty polyethylene sheeting or other approved covering material to retain moisture necessary for rapid curing inherent in high void surface areas.

  - Roll the polyethylene sheeting on tubes prior to pavement to allow the rolling over the fresh pavement across the width of the forms. Maintain a minimum of 300 mm (12 in.) overhang on each side of the form and use this excess sheeting to secure it in place.

  - Prior to covering, an evaporation retarder, such as soy bean oil, may be sprayed onto the pavement from both sides of the paving operation. Follow manufacturer's recommendations for application rate.

  - Tape or otherwise repair any holes, tears, or cuts in the plastic sheeting to prevent moisture loss and air infiltration under it.

  - Use anchors, such as boards and sand bags, to properly secure the edges of the sheeting along the pavement edge. Use anchors so plastic sheeting will remain securely in place for the entire curing period and not be removed by wind. Mud clumps, construction trash, rocks, subgrade material, etc. shall not be used as anchors. Prevent plastic sheeting from billowing in the wind during the entire 7-day cure. If using wood forms, staple the plastic sheeting outside of the forms. Place wooden boards on the upper outer edge for added continuous anchoring, and place sand bags every 1 to 2 m (3 to 6 ft) to hold them in place.

  - Secure the curing cover in place and do not move for a minimum of 7 days.

  - Do not permit vehicular traffic on the pavement until curing is complete and do not permit truck traffic for at least 14 days.

Figures 3-19 through 3-22 show typical placement and finishing activities.

Figure 3-19
Concrete deposited as close as possible to the final location. Finishing performed with a hydraulic roller screed, Overland Park, KS
Source: Concrete Promotional Group, Inc.

Figure 3-20
Finishing performed with a power-roller screed, Legacy Recreational Trail, Lexington, KY
Source: Concrete Promotional Group, Inc.

Figure 3-21
Industrial loading dock. Finishing performed with a pervious concrete laser screed, Lacey, WA
Source: Puget Sound Concrete Specification Council

Figure 3-22
Plastic sheeting secured on the pervious concrete, Overland Park, KS
Source: Concrete Promotional Group, Inc.

3.5   Post-construction Operation and Maintenance

See **Chapter 1 and Chapter 8** for surface maintenance and cleaning guidance. The most common complications related to pervious concrete are surface raveling (almost always caused by a dry mixture or not enough water), too much cementitious paste or improper curing, or clogging from excess soil deposited from adjacent unstabilized surfaces soon after construction. As shown in **Figure 3-23**, erosion and sediment controls were removed too early at this project site causing sediment-laden stormwater runoff to temporarily clog the pervious concrete.

Figure 3-23
Erosion and sediment control failure, GA
Source: WR Grace

Pervious concrete should be repaired with a visually and functionally similar pervious concrete mixture. Very thin repairs are inadequate and full-depth patches are required in most situations. Widespread surface raveling can often be mitigated by grinding the surface to remove the weak layer, followed by pressure washing and vacuuming to remove any debris. For any overlay, the underlying substrate should be clean and dampened just prior to overlay placement.

3.6   Cost Information

The overall cost of a pervious concrete pavement system varies significantly depending upon the particular application and market. The pervious concrete material alone generally costs about $21.50 to $64.50/m² ($2 to $6/sf) depending on the concrete thickness, size of the installation, and market. The typical costs installed are about $64.50 to $107.50/m² ($6 to $10/sf). As with any site practice, storm drainage, soils, and slope considerations all impact the overall installation costs of the pavement.

A cost comparison between pervious concrete pavement and other conventional types of pavement depends mostly on the structural and storm drainage requirements of the project. **Appendix B** includes a cost comparison table for the permeable pavement surface materials. Generally, the cost for pervious concrete is higher than the cost for traditional concrete, but results in savings when the storm drainage requirements associated with normal concrete cost more than the incremental increase for pervious concrete. When comparing the costs for a traditional infrastructure project and one with pervious concrete, the costs between the two options in the context of the entire project should be considered. Overall, the maintenance costs associated with pervious concrete are lower than those costs for impervious pavements since they reduce or eliminate the need for catch basins, pipes, ditches, and/or detention basins, all of which require maintenance.

3.7   Stormwater Benefits

Pollutant Removal

Like other permeable pavements, pervious concrete systems are effective in reducing certain pollutants, such as total suspended solids (TSS), metals, oil, and grease. Large particles are caught on the surface for vacuum cleaning. Small particles are caught in the aggregate base. Oils and grease adsorb to the concrete and aggregate. They are reduced through evaporation, UV degradation, and microbial action. The immediate 0.6 m (2 ft) of underlying natural soil is biologically-active, helping to reduce oils, greases, and other pollutants such as nutrients associated with fertilizers.

Heat Island Effect

Pervious concrete can help reduce the heat island effect through at least two mechanisms. The open structure allows evaporative cooling from moisture contained in the pervious concrete layer and underlying layers. The void structure of pervious concrete and the aggregate storage layer may reduce the depth of heating from the surface. Reduced heating and lower mass than traditional pavements result in less energy stored and released later. Research shows that a pervious concrete system (pervious concrete and aggregate base) with a measured solar reflectance index (SRI) of 14 stores less energy than a conventional concrete pavement with a SRI of 37 (Haselbach 2011). Because of this research, the International Green Construction Code (IGCC) applies pervious concrete as a heat island mitigation option independent of surface color (IGCC 2012).

3.8   Resources for Specifications

Specifications for pervious concrete have become increasingly available in recent years through industry and technical groups. As with all permeable pavements, each specification must be developed for site and local conditions. No single specification is directly applicable to the entire United States without some modifications. The following associations/groups/boards are sources for pervious concrete specification guidance. Sources include:

- American Concrete Institute (ACI) 522 Pervious Concrete Committee
- The National Concrete Pavement Technology Center (CP Tech Center)
- National Ready Mixed Concrete Association (NRMCA)
- American Society for Testing and Materials (ASTM) C09.49 Pervious Concrete Committee
- Transportation Research Board (TRB)
- Local ready mixed concrete association or NRMCA concrete sponsoring groups

3.9   General Resources

Helpful resources are provided as follows:

GENERAL DESIGN GUIDANCE

American Concrete Institute (ACI) (2010). "522-R10: ACI 522 Committee Report." *Pervious Concrete*, Farmington Hills, MI.

Delatte, N., Miller, D., and Mrkajic, M. (2007). "Portland Cement Pervious Concrete: Field Performance Investigation on Parking Lot and Roadway Pavements." *Final Report of the RMC Research and Education Foundation*, Silver Springs, MD.

Ferguson, B. K. (2005). *Porous Pavements*. Taylor and Francis Group. New York, NY.

Haselbach, L. (2010). *The Engineering Guide to LEED—New Construction, Sustainable Construction for Engineers*. New York: McGraw-Hill Professional.

Haselbach, L., Valavala, S., and Montes, F. (2006). "Permeability Predictions for Sand Clogged Portland Cement Pervious Concrete Pavement Systems," *Journal of Environmental Management*, 81, 42-49.

Tennis, P. D., Leming, M. L., and Akers, D. J. (2004). *Pervious Concrete Pavements*, EB302, Portland Cement Association, Skokie, Illinois, National Ready Mixed Concrete Association, Silver Spring, MD.

Transportation Research Board (TRB) (2010). *Transportation Research Record: Journal of the Transportation Research Board*, No. 2164, Transportation Research Board of the National Academies, Washington D.C., 82-88, DOI 10.3141/2164-11.

United Facilities Guide Specifications (UFGS) (2010). "Pervious Portland Cement Concrete for Roads and Site Facilities," Section 32 13 13.06, 21, <www.wbdg.org>.

HYDROLOGIC DESIGN

Leming, M., Malcom, H.R., and Tennis, P. (2007). "Hydrologic Design of Pervious Concrete." Portland Cement Association, EB303, Skokie, IL.

Portland Cement Association (PCA) (2007). "Pervious Concrete Hydrological Design and Resources." (CD-ROM), Portland Cement Association, CD063.02, Skokie, IL.

STRUCTURAL DESIGN

Delatte, N. (2007). "Structural Design of Pervious Concrete Pavement." *Proc., Transportation Research Board (TRB) Annual Meeting*, Washington, D.C.

Goede, W. and Haselbach, L. (2011). "Investigation into the Structural Performance of Pervious Concrete." *ASCE Journal of Transportation Engineering*, doi:10.1061/(ASCE)TE.1943-5436.0000305.

Kevern, J. T., Wang, K., and Schaefer, V. R. (2009a). "Evaluation of Pervious Concrete Workability Using Gyratory Compaction." *American Society of Civil Engineers Journal of Materials in Civil Engineering*, 21(12).

Kevern, J. T., Wang, K., and Schaefer, V. R. (2009b). "Test Methods for Characterizing Air Void Systems in Portland Cement Pervious Concrete." *Journal of ASTM International,* Selected Technical Papers STP 1511 Recent Advancements in Concrete Freezing and Thawing (FT) Durability, 6(9).

Kevern, J. T., Wang, K., and Schaefer, V. R. (2009c) "The Effect of Curing Regime on Pervious Concrete Abrasion Resistance." *Journal of Testing and Evaluation*, 7(4), JTE101761.

Kevern, J. T., Wang, K., and Schaefer, V. R. (2010). "The Effect of Coarse Aggregate on the Freeze-Thaw Durability of Pervious Concrete." *American Society of Civil Engineers Journal of Materials in Civil Engineering*, 22(5).

Rodden, R., Voigt, G., and Gieraltowski, A. (2010). "PerviousPave." (Software), American Concrete Paving Association (ACPA), Skokie, IL.

MIXTURE PROPORTIONING

Kevern, J. T., Schaefer, V. R., Wang, K., and Suleiman, M. T. (2008). "Pervious Concrete Mixture Proportions for Improved Freeze-Thaw Durability." *Journal of ASTM International*, 5(2), doi:10.1520/JAI101320.

Schaefer, V. R., Wang, K., Sulieman, M. T., and Kevern, J. (2009). "Mix Design Development for Pervious Concrete in Cold Weather Climates." National Concrete Pavement Technology Center (CP Tech Center), Iowa State University, Ames, IA, <http://www.ctre.iastate.edu/reports/mix_design_pervious.pdf>.

SPECIFICATIONS

American Concrete Institute (ACI) (2008). "Specification for Pervious Concrete Pavement." ACI Standard 522.1-08, Farmington Hills, MI.

Kevern, J., and Zufelt, J. (2013). "Introduction to ASCE Monograph-Permeable Pavements in Cold Climates." *ISCORD 2013: Planning for Sustainable Cold Regions*, Zufeit, J. E. ed., Amercian Society of Civil Engineers, 482-494, doi: 10.1061/9780784412978.047 <http://dx.doi.org/10.1061/9780784412978.047>.

United Facilities Guide Specifications (UFGS) (2010). "Pervious Portland Cement Concrete for Roads and Site Facilities," Section 32 13 13.06, 21, <www.wbdg.org>.

SPECIFIER'S GUIDES

Colorado Ready Mixed Concrete Association (CRMCA) (2010). "Specifier's Guide to Pervious Concrete Pavement Design," 1(2), Centennial, CO, <www.crmca.org>.

Concrete Promotional Group (CPG) MO/KS American Concrete Pavement Association (2009). "Specifier's Guide to Pervious Concrete Pavement in the Greater Kansas City Area." Overland Park, KS, <www.concretepromotion.com>.

Ohio Ready Mixed Concrete Association (ORMCA) (2009). "Specifier's Guide to Pervious Concrete Pavement with Detention," *Ohio Concrete*, <www.ohioconcrete.org> (October, 2009).

CONSTRUCTION

National Ready Mixed Concrete Association (NRMCA) (2010). "Text Reference for Pervious Concrete Contractor Certification," NRMCA Publication 2PPCRT, Silver Springs, MD.

Offenberg, M. (2005). "Producing Pervious Pavements." *Concrete International*. Farmington Hills, MI. 27(3), 50–54.

3.10 References

American Concrete Institute (ACI) (2008). "Specification for Pervious Concrete Pavement," ACI Standard 522.1-08, Farmington Hills, MI.

American Concrete Institute (ACI) (2010). "Pervious Concrete 522-R10 ACI 522 Committee Report." Farmington Hills, MI.

American Society for Testing and Materials (ASTM) (2008). ASTM C-1688 "Standard Test Method for Density and Void Content of Freshly Mixed Pervious Concrete." *Annual Book of ASTM Standards*, Vol. 4.02, West Conshohocken, PA.

American Society for Testing and Materials (ASTM) (2009). ASTM C-1701 "Standard Test Method for Infiltration Rate of In Place Pervious Concrete." *Annual Book of ASTM Standards,* 4(2), West Conshohocken, PA.

American Society for Testing and Materials (ASTM) (2011). ASTM C-1747 "Standard Test Method for Determining Potential Resistance to Degradation of Pervious Concrete by Impact and Abrasion." *Annual Book of ASTM Standards,* 4(2), West Conshohocken, PA: ASTM International.

American Society for Testing and Materials (ASTM) (2012). ASTM C-1754 "Standard Test Method for Density and Void Content of Hardened Pervious Concrete." *Annual Book of ASTM Standards,* 4(2), West Conshohocken, PA: ASTM International.

Delatte, N. (2007). "Structural Design of Pervious Concrete Pavement." *Proc., Transportation Research Board (TRB) Annual Meeting*, TRB 07-596, Washington, D.C.

Goede, W. and Haselbach, L. (2011). "Investigation into the Structural Performance of Pervious Concrete."

American Society of Civil Engineers Journal of Transportation Engineering, 10.1061/(ASCE)TE,1943-5436.0000305.

Haselbach, L., Boyer, M., Kevern, J., and Schaefer, V., (2011), Cyclic Heat Island Impacts in Traditional versus Pervious Concrete Pavement Systems." *Transportation Research Record: Journal of the Transportation Research Board (TRB)*, No.2240, 107-115.

The International Green Construction Code (IGCC) (2012). International Code Council (IOCC), Washington D.C.

Kevern, J. T., (2011). "Design of Pervious Concrete Mixtures." (CD-ROM), Design software distributed by the National Pervious Concrete Paving Association (NPCPA), V3.0 <http://www.npcpa.org/>.

Kevern, J. T., Biddle, D., and Cao, Q. (2013a). "Effect of Macro-Synthetic Fibers on Pervious Concrete Properties." *American Society of Civil Engineers Journal of Materials in Civil Engineering*, submitted Oct. 15, 2013.

Kevern, J. T. and Sparks, J. D. (2013b). "Low Cost Techniques for Improving the Surface Durability of Pervious Concrete." *Transportation Research Record: Journal of the Transportation Research Board (TRB)*, Concrete Materials, No. 2342, Transportation Research Board of the National Academies, Washington D.C., 83-89.

Kevern, J. T., Biddle, D., and Cao, Q., (2014) "Effect of Macro-Synthetic Fibers on Pervious Concrete Properties." *ASCE Journal of Materials in Civil Engineering*, accepted Jan. 07, 2014.

Leming, M., Malcom, H.R., and Tennis, P. (2007). "Hydrologic Design of Pervious Concrete." Portland Cement Association, EB303, Skokie, IL.

National Ready Mixed Concrete Association (NRMCA) (2010). "Text Reference for Pervious Concrete Contractor Certification." *NRMCA Publication* 2PPCRT, Silver Springs, MD.

Portland Cement Association (PCA) (2007). "Pervious Concrete Hydrological Design and Resources." (CD-ROM), Portland Cement Association, CD063.02, Skokie, IL.

Rodden, R., Voigt, G., and Gieraltowski, A. (2010a). "Structural and Hydrological Design of Pervious Concrete Pavements using PerviousPave." *Proc., FHWA International Conference on Sustainable Concrete Pavements,* Sacramento, CA, 8.

Rodden, R., Voigt, G., and Gieraltowski, A. (2010b). "PerviousPave." (Software), American Concrete Paving Association (ACPA), Skokie, IL.

Schaefer, V. R., Wang, K., Sulieman, M.T., and Kevern, J. (2006). "Mix Design Development for Pervious Concrete in Cold Weather Climates." National Concrete Pavement Technology Center (CP Tech Center), Iowa State University, Ames, IA, <http://www.ctre.iastate.edu/reports/mix_design_pervious.pdf>.

Schaefer, V. R., Kevern, J. T., Izevbekhai, B., Wang, K., Cutler, H., and Wiegand, P. (2010). "Construction and Performance of the Pervious Concrete Overlay at MnROAD."

Suleiman, M., Gopalakrishnan, R., and Kevern, J. T. (2011). "Structural Response of Pervious Concrete Pavement Systems Using Falling Weight Deflectometer Testing and Analysis." *American Society of Civil Engineers Journal of Transportation in Civil Engineering*, doi:10.1061/(ASCE)TE.1943-5436.0000295 <http://dx.doi.org/10.1061/(ASCE)TE.1943-5436.0000295>.

Permeable Interlocking Concrete Pavement (PICP) Fact Sheet

DESCRIPTION

Permeable interlocking concrete pavement (PICP) consists of manufactured concrete units that reduce stormwater runoff volume, rate, and pollutants. The impervious units are designed with small permeable joints. The openings typically comprise 5% to 15% of the paver surface area and that maintain high permeability with small-sized aggregate fill. The joints allow stormwater to flow into a crushed stone aggregate bedding layer and base/subbase that support the pavers, while providing water storage as well as runoff quantity and quality treatment. PICP is visually attractive, durable, easily repaired, requires low maintenance, and can withstand heavy vehicle loads.

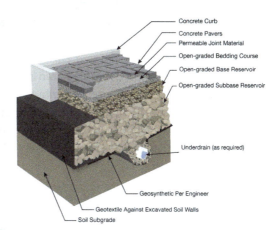

Source: ICPI 2012

APPLICATIONS

POTENTIAL APPLICATION		NOTES
Overflow Parking	Yes	
Primary Parking Areas (most heavily used)	Yes	
Sidewalks/Pathways	Yes	
Drive/Aisles	Yes	
Roads/Highways	Limited	Can be used for shoulders, rest areas, etc. up to design load repetition limitations
Access Drives/Ring Roads	Yes	Up to design load repetition limitations
Loading Areas	Yes	Up to design load repetition limitations
Frequent Truck Traffic	Yes	Up to design load repetition limitations

Permeable Interlocking Concrete Pavement (PICP) Fact Sheet continued

BENEFITS

PICP requires no curing time and is ready for traffic upon installation. It accommodates soil infiltration rates as low as 1.3 mm/hr (0.05 in./hr). High stormwater storage, treatment, and increased structural capacity are achieved with thicker subbases. PICP supports AASHTO H20 loads and lifetime design repetitions up to 1 million; 18,000 lb (80 kN) equivalent single axle loads (ESALs) or Caltrans Traffic Index (TI) of 9. Higher lifetime load repetitions can be achieved with stabilized bases. Paving units can be colored to mark driving/parking lanes and striping. Light colored pavers contribute to albedo (reflectance) and reduced urban heat island. Paving units can include TiO_2 photocatalytic cement/pigment to reduce smog and increase surface albedo. PICP can be combined with horizontal ground source heat pumps to reduce building heating and cooling costs. It maintains infiltration with sediment in the surface openings, which can be (vacuum) cleaned to restore surface infiltration.

STORMWATER QUANTITY FUNCTIONS		NOTES
Volume	High	Volume control, increases with additional storage provided in the aggregate base/subbase
Groundwater Recharge	High	Recharge high if soils permit
Peak Rate	Medium/ High	Peak discharge rate control increased with increased storage and/or high soil permeability
STORMWATER QUALITY FUNCTIONS		NOTES
Total Suspended Solids	High	Pre-treatment encouraged but not required; >80% TSS reduction possible without pretreatment
Total Nitrogen/ Total Phosphorus	Moderate	High nutrient quantity reductions when detaining water in base/subbase for at least 24 hours
Metals	High	
Temperature	High	Maintains and does not increase ambient rainfall temperatures

SITE CONSTRAINTS/CONCERNS

- Limit runoff from contributing impervious surfaces to no greater than five times the PICP area
- Feasibility on steeper slopes possible with design modifications such as bermed/terraced soil subgrade to control down slope flows
- Low permeability soils may require use of perforated underdrains. Maintain the bottom of the pavement base no less than 0.6 m (2 ft) above the seasonal high water table. Some situations may allow the design of perforated drain pipes to remove water, should it rise above this depth. Systems with impermeable liners should not allow the seasonal high water table within less than 0.3 m (1 ft) from the bottom of the liner. Protection and vacuum sweeping are required to maintain clean aggregates and finished surfaces during construction.
- Geotextile use for soil separation and soil compaction are at the design engineer's discretion.

Source: © ICPI

Permeable Interlocking Concrete Pavement (PICP) Fact Sheet continued

RECOMMENDED KEY DESIGN CRITERIA

- Complete soil subgrade infiltration testing.
- Maintain uncompacted, permeable soil subgrades for infiltration, where possible.
- Provide stationary edge restraints at PICP perimeter to maintain interlock among paving units.
- Stone base layer remains 100 mm (4 in.) thick. Subbase thickness varies according to anticipated traffic loads, soil subgrade support, and required stormwater storage. A subbase layer is not required in pedestrian or residential driveway applications; use base layer only at appropriate thickness for water storage and to support loads.
- Provide overflow at lowest end(s) via pipes or through surface.
- Use 80 mm (3.15 in.) thick pavers conforming to ASTM C936 in areas subject to vehicular traffic.
- AASHTO minimum layer coefficient of 0.3 for the paver and bedding layer.

PERFORMANCE

HIGH—55 MPa (8,000 psi) compressive strength concrete units per ASTM C936 provide high resistance to loads, freeze-thaw, and deicing salts. Modular surface accommodates underground utility repairs without cutting, damaging and decreasing pavement life. Surface openings between pavers can be cleaned via vacuumed, if clogged.

OPERATION & MAINTENANCE

MEDIUM—Vacuum one (1) to two (2) times annually with a regenerative air vacuum sweeper plus any adjacent impervious surfaces. Adjust cleaning schedule as required. If surface cleaning is neglected for years, use true vacuum sweeper to remove encrusted sediment from openings, then refill openings with void and joint aggregate to restore higher surface infiltration. Inspect annually; top up voids and jointing materials as required. Use jointing materials for tire traction in winter. The use of sand discouraged. PICP can process oil drippings from vehicles. The surface is ADA compliant.

COSTS

MEDIUM—For minimum 1,500 m² (15,000 sf) area mechanically installed using prevailing labor wages with 80 mm (3.15 in.) thick pavers (mechanically installed), jointing and bedding materials: $0.30 to $0.40/m² ($3 to $4/sf) (2012 cost estimate).

REFERENCES

United States Environmental Protection Agency (US EPA) (2009). "Permeable Interlocking Concrete Pavement." National Pollutant Discharge Elimination System (NPDES), <http://water.epa.gov/polwaste/npdes/swbmp/Permeable-Interlocking-Concrete-Pavement.cfm>.

4 Permeable Interlocking Concrete Pavement (PICP)

4.1 System Description

Permeable interlocking concrete pavement (PICP) consists of (impervious) manufactured concrete units that form permeable voids and joints, when assembled into a laying pattern. The openings typically comprise 5% to 15% of the paver surface area that maintain high permeability with small-sized aggregates. The openings allow stormwater to enter a permeable stone bedding layer and base/subbase that support the pavers while providing storage and runoff treatment. PICP replaces traditional impervious pavement for most pedestrian and vehicular applications, except high-volume/high-speed roadways. PICP has performed successfully in pedestrian walkways, sidewalks, driveways, parking lots, and low-volume roadways. Key resources for designers are published by the Interlocking Concrete Pavement Institute (ICPI), which include a detailed manual entitled *Permeable Interlocking Concrete Pavements: Selection Design Construction Maintenance* (Smith 2011) and a design software program entitled *Permeable Design Pro* (ICPI 2009). See Chapter 9 for more modeling resource software and tools.

The paving units are made of manufactured concrete that conforms to ASTM C936 *Standard Specification for Solid Concrete Interlocking Paving Units* (ASTM 2013) in the United States and Canadian Standards Association A231.2 *Precast Concrete Pavers* (CSA 2014) in Canada. The ASTM product standard requires an average compressive strength of 55 MPa (8,000 psi), less than 5% absorption, and freeze-thaw as well as abrasion durability requirements.

The paving units are made of highly durable concrete, resistant to freeze-thaw and the deleterious effects of deicing salts. The units are manufactured in different proprietary and non-proprietary shapes that render a range of laying patterns, joints, and openings. Various colors enable pavement markings, plus architectural context sensitivity. In addition, light colored units can be manufactured to increase reflectivity and the solar reflective index. **Figure 4-1** shows an application in Illinois where 1.6 km (1 mile) of PICP was placed. The PICP installation was credited with eliminating local flooding due to a formerly inadequate storm drainage system.

Figure 4-1
PICP roadway, Main Street, Warrenville, IL
Source: ICPI

Several benefits associated with PICP are listed. Some of these are common to other permeable pavements. Stormwater specific benefits are described in more detail later in this chapter.

Paver Surface/Units

- 50-year life
- ADA compliant
- Colored units can mark parking stalls and driving lanes; light colors can reduce night time lighting needs
- Capable of plowing with conventional snow removal equipment
- Durable, high-strength, and low-absorption concrete units resist freeze-thaw, heaving, and degradation from deicing materials
- Various paver shapes and designs are offered by the industry, often with distinct characteristics
- Can provide traffic calming
- Reduces puddles on parking lots
- Pavers can include photocatalytic materials to reduce air pollution
- High solar reflectance surface helps reduce micro-climatic temperatures and contributes to urban heat island reduction
- Units manufactured with recycled materials and cement substitutes that helps to reduce greenhouse gas emissions

Construction

- Immediately ready for traffic upon completion; no additional days needed for curing
- Can be installed in freezing weather if subgrade and aggregates remain unfrozen
- Capable of wet weather (light rain) installation
- No time-sensitive materials that require site forming and management for curing
- Contractor training and credentials available through the Interlocking Concrete Pavement Institute (ICPI)
- Sites where space constraints, high land prices, and/or runoff from additional development make PICP a cost-effective solution

Maintenance & Repairs

- Paving units and base materials can be removed and reinstated
- Utility cuts do not damage the surface and decrease pavement life; no unsightly patches from utility cuts
- Capable of winter maintenance and repairs
- Reduced need of deicing materials during winter season due to higher heat exchange with the soil subgrade and rapid ice melt
- Reduced liability from slipping on ice due to rapid ice melt and surface infiltration
- Surface cleaning with standard vacuum equipment
- Highly clogged surfaces can be restored with vacuum equipment

4.2   Applications—Site Constraints/Concerns

Applications and General Site Selection Criteria

PICP can be used in the following applications:

- Walks, parking lots, and main and service drives around commercial, institutional, recreational, and cultural buildings
- Boat ramps
- Industrial sites that do not receive hazardous materials (i.e., where there is no risk to groundwater or soils from spills)

Site Considerations:

- The impervious area should not exceed five times the area of the PICP receiving the runoff.
- The pavement should be located down slope from building foundations, and any nearby foundations should have piped drainage at the footers. Basement walls in close proximity to PICP systems may require waterproofing, which may include an impermeable liner placed vertically against them.
- A typical surface slopes less than 5%. The use of subgrade berms or barriers to reduce lateral flows and promote infiltration can allow for applications on steeper surface slopes of up to 15%.
- Land surrounding and draining into the pavement should not exceed 20% slope. The surrounding land use draining to the PICP surface must be stabilized and, preferably, impervious.
- The property owner must be able to meet maintenance requirements to provide the long-term function of the pavement system.
- There should be no post-construction increase in impervious cover draining into the pavement, unless the pavement is designed to infiltrate and store runoff from future increases in impervious cover.

Climate Considerations

Many years of experience and monitoring have demonstrated that PICP does not heave when frozen. This is evidenced by many PICP projects in Chicago, Minneapolis, and Toronto remaining stable during freezing and thawing climates. This is due to the following factors:

- The pavement base/subbase and saturated soil subgrade drains prior to freezing.
- The air in the aggregate voids provides some insulating effect in slowing the movement of freezing temperatures toward the soil subgrade.
- The earth provides some heat to delay freezing of the soil subgrade and base/subbase such that both can drain prior to freezing.
- Should water freeze in the base or subbase, there is sufficient space in the aggregate voids for the frozen water to expand as it freezes.
- A benefit of all permeable pavements is that once snow is plowed, the remaining snow can melt and infiltrate into the surface when temperatures rise, thereby reducing or eliminating re-freezing at night and ice hazards. This condition also reduces the need for deicing salts.

Snow removal considerations are discussed in **Section 4.5 Post-construction Operation and Maintenance** of this chapter.

4.3   General Design Criteria—General Specifications

General Design Considerations

If intended to achieve significant stormwater infiltration, PICP should follow the general design guidelines in **Chapter 1** and the following design criteria:

As with all permeable pavements, PICP can be designed for full-, partial- or no-infiltration of water into the soil subgrade. For full- or partial-infiltration designs, it is good practice, and often a regulatory requirement, to have at least 0.6 m (2 ft) of soil between the bottom of the subbase and the elevation of the seasonal high water table for pollutant filtering purposes as well as soil stability when saturated. For no-infiltration designs, this height can be reduced to 0.3 m (1 ft) under the PICP subbase. A number of full- and partial-infiltration PICP projects prove to successfully perform in coastal areas. Many projects are built over sandy soils and high water tables with tidal influences in groundwater levels. Sandy soils offer stability even when saturated, and underdrains can be designed to remove or reduce high groundwater levels within the PICP base/subbase. There is always design consideration given to overflow conditions since all permeable pavements are designed to infiltrate and drain water volumes from a specific range of storms. Excess water from high depth rainfall or that from repeated storms is best handled through drain pipes at the lowest elevations of the PICP. Preferred drainage methods allow overflow from extreme rainfall events to be directed into downstream stormwater BMPs, conveyance systes (e.g. stormdrains) or receiving waters. Drainage methods otherwise would allow excess water to rise out of and drain from the surface, which could potentially mobilize sediment and pollutants situated in the stone-filled PICP openings.

Structural Design Considerations

PICP structural design relies on the flexible pavement AASHTO equation in the AASHTO (1993) *Guide for Design of Pavement Structures*. This method calculates a Structural Number (SN) given traffic loads, soil type, climatic, and soil moisture conditions. The designer then finds the appropriate combination of pavement surfacing and base materials, whose strengths are characterized with layer coefficients. When added together, the coefficients (representing various pavement material thicknesses) should meet or exceed the SN required for a design. This empirical design approach appears to be applicable to PICP with some consideration given to layer coefficients.

PICP structural design assumes a soil California Bearing Ratio (CBR) (minimum 96-hour soaked per ASTM D1883 or AASHTO T193) strength of at least 4%, an R-value of 9, or a resilient modulus (M_r) over 40 MPa (6,000 psi) to qualify for use under vehicular traffic. The compaction required to achieve this will greatly reduce the infiltration rate of the soil. Therefore, the permeability or infiltration rate of soil should be assessed at the density required to achieve at least 4% CBR. PICP can be placed on lower strength soils, but may require thicker and/or cement-stabilized subbases.

Base/Subbase Layer Coefficients

As noted above, a key input for flexible pavement design is the layer coefficient that characterizes the stiffness of each pavement layer. The numerical layer coefficient is expressed per inch or per millimeter of pavement layer thickness. The thickness of each material is multiplied by the layer coefficients and all coefficients are added to see if they equal or exceed the SN to satisfy the design.

Unlike dense-graded bases, there is no established AASHTO layer coefficient for open-graded bases. A conservative approach is recommended and more research on this is needed. For design purposes, layer coefficients should be less than that used for design of dense-graded bases (typically 0.12 to 0.14). Layer coefficients of 0.06 to 0.10 for open-graded bases and subbases are recommended.

Paver and Bedding Layer Coefficients

A conservative recommendation for the AASHTO layer coefficient for the paver and bedding layer is 0.3. Typically, concrete pavers have a thickness of 80 mm (3.15 in.) and a bedding thickness of 50 mm (2 in.) for the ASTM No. 8 stone. This yields a SN for this layer of 1.54 (or 5.125 x 0.3). Concrete paver manufacturers should be consulted for additional structural design data for specific paver shapes and laying patterns. ICPI recommends that PICP subject to vehicular traffic be placed in a 90 or 45 degree herringbone pattern.

Hydrologic Design

The hydrologic design of PICP is similar to that of other types of permeable pavements (see **Chapter 1 and Chapter 9**).

Permeable Interlocking Concrete Materials

Paver surface/units

The following requirements/data are needed for paver design:

Figure 4-2
Various types of paving units used in PICP
Source: ICPI

- Minimum thickness: 80 mm (3.15 in.) for vehicular applications and 60 mm (2.36 in.) for pedestrian applications. Paver voids and joints should be filled with stone to the bottom of paver chamfers to comply with the American with Disabilities Act (ADA) Design Guidelines (2010).

- Test results should indicate conformance to ASTM C936 *Standard Specification for Solid Interlocking Concrete Paving Units* or CSA A231.2 *Precast Concrete Pavers* as appropriate to the project location. If the dimensions of the units are larger than those stated in these standards, then CSA A231.1 *Precast Concrete Paving Slabs* is recommended as a product standard. Paving slabs should be used in pedestrian applications only. **Figure 4-2** illustrates many concrete paver shapes and patterns that can be utilized in PICP.

Materials for the Base, Bedding and Openings

The following data are required on all aggregate materials (i.e., base and subbase, bedding course, and aggregate in the pavement openings):

- Sieve analysis, including washed gradations per ASTM C136

- Void content in percent for the open-graded base per ASTM C29. This test method provides an approximate void space percentage, but does not represent in-place compacted density and resulting void content or porosity.

- Crushed stone, open-graded subbase and base—this material should be a hard, durable rock with 90% fractured faces and a Los Angeles (LA) Abrasion of <40 per ASTM C535 and C131 when subject to vehicular traffic. A minimum effective porosity of 0.32 per ASTM C29.

Besides use as a bedding material, ASTM No. 8 crushed stone aggregate is also recommended for filling the paver joints. Smaller sized aggregate such as No. 89 or No. 9 may be needed for joint filling material for paving units with narrow joints.

The base is consistently 100 mm (4 in.) thick and acts as a choker course between the bedding layer and the larger aggregate subbase beneath. Sometimes this layer is called a "choke course." The base layer is typically ASTM No. 57 stone or similar sized aggregate. For pedestrian applications and residential driveways, No. 57 can be used for the entire base and subbase. In such cases, the minimum thickness will be 150 mm (6 in.), but is often thicker for water storage and infiltration.

High performance aggregates coated with chemicals can reduce nutrients such as phosphorus and nitrogen. The chemical coatings have a limited time frame of removal effectiveness, typically less than 10 years. See **Chapter 6** for additional information on these treated aggregates.

Ferguson (2005) provides filter criteria for all aggregate layers and these are noted in **Table 4-1** below. D_x is the particle size at which x percent of the particles are finer as measured by mass. For example, D_{15} is the particle size of the aggregate for which 15% of the particles are smaller and 85% are larger.

Table 4-1 Filter Criteria for PICP Bedding, Base, and Subbase Aggregates

Permeability	D_{15} Base/D_{15} Bedding layer > 5
Choke	D_{50} Base/D_{50} Bedding layer < 25
Choke	D_{15} Base/D_{85} Bedding layer < 5

Source: ICPI

Subbase/Reservoir Course

For PICP in vehicular traffic, the aggregate subbase under the base consists of ASTM No. 2 crushed stone. This layer is placed without geotextile in many applications. This subbase material ranges in size from 19 to 63 mm (0.75 to 2.5 in.) and provides a low stress-dependent, stable working platform for construction equipment to spread and compact the No. 57 stone base. After compacting the No. 2 stone, No. 57 stone is spread and compacted or choked into the openings of the No. 2 stone.

A nuclear density gauge can only be used in backscatter mode to check the density of the No. 57 (or similar gradation) base layer. Such gauges are ineffective in measuring density of aggregate subbases. More effective devices are available to measure compacted stiffness, rather than density, of the base and subbase layers as well as the soil subgrade at known moisture contents; these include lightweight deflectometers and soil stiffness gauges. The use of lightweight deflectometers should comply with

ASTM E2583 *Standard Test Method for Measuring Deflections with a Light Weight Deflectometer* (LWD). The use of soil stiffness gauges should comply with ASTM D6758 *Standard Test Method for Measuring Stiffness and Apparent Modulus of Soil and Soil-Aggregate In-Place by Electro-Mechanical Method.*

The 100 mm (4 in.) thick base and subbase layers together provide the combined reservoir course for water storage. If ASTM No. 2 is not available for the subbase, ASTM or AASHTO No. 3 or No. 4 gradations can be used. These materials have approximately 40% void space. Subbases and bases should be uniformly compacted such that there is no visible movement after static rolling with a 10T (10 ton) roller.

Subbases exposed to vehicular traffic will likely exceed 200 mm (8 in.) thickness when the soil subgrade is weak (CBR < 4% or R-value < 9), has high amounts of clay or silt, or subject to frost heaving. Likewise, thicker subbases or those stabilized with cement, may be required over a seasonally high water table that creates an unstable soil subgrade in low-lying areas subject to flooding or over continually saturated soils. Perforated drain pipes at the top of the soil subgrade can provide additional stability by helping to avoid saturated conditions that might lead to premature rutting. PICP manufacturers' literature may have further recommendations on base thickness and materials as a function of the soil strength and anticipated vehicular loads.

Some state transportation agencies specify well-graded aggregates as permeable bases (drainage layers) under conventional pavements, which are either unstabilized or stabilized with asphalt or cement). These same aggregates can be used in PICP as long as they meet certain gradations, and only under certain conditions, as detailed below. In terms of gradation, these bases can have a higher percentage of material passing the 4.75 mm (0.19 in.) (No. 4) sieve than those described in this chapter, but are limited to 2% passing the 0.075 mm (0.001 in.) (No. 200) sieve. Using such bases for PICP can be a design option when water storage is not the primary objective—for example, when smaller rainfall depths are captured, treated, and released. Stability is increased by using underdrains that prevent saturation of the aggregates and related pore pressures. Such underdrains are recommended in all applications to prevent this condition. Consultation with an experienced pavement engineer is recommended regarding assessment of the structural contribution of such bases especially when stabilized with asphalt or cement. Other aggregate bases may also be used if they are deemed to meet the hydrological and structural requirements of the projects by the engineer/designer of record.

4.4   Recommended Installation Guidelines

ICPI offers an educational program called the PICP Installer Specialist Course that issues a record of completion to participants. Project specifications should include verification of a PICP Installer Specialist Course record of completion and project/work history at the time of bid document submission by the installation subcontractor/general contractor. Construction specifications should state that the PICP installation crew shall employ at least one person holding a current ICPI PICP Installer Specialist Course record of completion who must be on site to function as project foreman, overseeing all crews during all PICP installations. In addition, it is strongly recommended that the PICP project inspectors hold a PICP Installer Specialist Course record of completion in order to facilitate informed inspection. Inspection is essential on areas subject to vehicular traffic such as parking lots and low-speed roads.

A pre-construction meeting is essential to the success of all PICP projects. The guide construction specification noted later in this section provides a list of items that need to be determined at the pre-construction meeting.

Construction Operations

Like all permeable pavements, PICP aggregates must not be contaminated with sediment during transfer, installation, and use during construction. The pavement should be built late into the construction schedule. If this is not possible, the subbase and base can be constructed with a layer of geotextile and aggregates installed to support construction traffic. The soiled aggregate and geotextiles are replaced with the bedding and pavers near the end of the project. Another approach, if the site allows, is building a temporary road for construction access that does not interfere with PICP construction, and instructing all trades to use the construction access road. Additional consideration should be given to installing a truck tire-washing station for projects that have a high risk of sedimentation. The guide construction specification at the end of this chapter provides additional considerations.

4.5   Post-construction Operation and Maintenance

Post-construction maintenance is essential to long-term PICP structural and infiltration performance. Like all permeable pavements, PICP surfaces can become clogged with sediment over time, thereby decreasing the surface infiltration rate. The rate of sedimentation depends on the amount of traffic and other sources that wash sediment into the voids, joints, base, and soil. Weeds should be removed, since they will affect surface infiltration. Weeding by hand is the safest alternative. Herbicides are not recommended since they can harm the environment.

All permeable pavements take on sediment and experience a reduction in their surface infiltration rate. This decrease can be as much as 80% to 90% if regular surface vacuuming is not done. Researchers (Bean 2004 and 2005; Gerrits 2002; Kinter 2010) and practical experience have demonstrated that periodic sediment removal from the openings using vacuum equipment increases surface infiltration rates. Vacuum sweeping (without water spray) using normal, regenerative air street cleaning equipment can remove loose sediment from the openings. Upon site inspection, vacuum sweeping should be done once or twice annually with regenerative air equipment. The surface should be monitored for more (or less) frequent cleaning. Some equipment has the potential to vacuum stones from the pavement openings so suction adjustments may be required.

In more severe cases where vacuuming has not been performed for years, true vacuum equipment can remove dirt trapped in the openings to restore surface filtration. Such equipment has twice the vacuum force as regenerative air equipment. True vacuum equipment can remove the first 13 to 25 mm (0.5 to 1 in.) of aggregate and dirt in the openings. On-site engine RPM adjustment of true vacuum equipment may be required so these machines can withdraw all material from the openings. The cleaned openings can then be replenished with clean aggregate, increasing surface infiltration. Research by Chopra (2010) examined the cleaning highly clogged permeable pavements with true vacuum equipment and evaluated results by using surface infiltration tests. PICP demonstrated good restorative characteristics for surface infiltration.

Snow can be plowed from pavers, as with any other pavement. Since deicing salts will infiltrate into the base and soil, they should be applied sparingly because they can accumulate in the soil subgrade. Research at the Toronto and Region Conservation Authority (Van Seters 2007) indicates that PICP requires less deicing salt than impervious asphalt pavements, since permeable pavement remains warmer throughout the winter by as much as 8° C (14° F). Sand should not be applied on the PICP for traction because it will accelerate surface clogging. If traction is required, No. 8 stone (or similar) should be used. If sand is used, the PICP surface should be vacuumed in the spring to reduce the risk of decreased surface infiltration.

Cracked or damaged units can be removed and replaced. The same units can be reinstated after repairs to the base, drain pipes, liners, or underground utilities have been made. Sealers can be applied to the pavers without overflow onto the aggregates by using a roller. Such sealers can enhance appearance, while making stains easier to remove.

PICP is sometimes constructed with an observation well. The well is typically a 100 to 150 mm (4 to 6 in.) diameter perforated pipe with a screw cap just slightly below the surface of the pavers that can be removed to observe the subgrade infiltration rate. The cap should lock and be vandal-resistant. The depth to the soil subgrade should be marked on the lid. The observation well is located in the furthest down-slope area within 1 m (3 ft) from the pavement edge. The top of the pipe can also be placed under the pavers. This hides the cap from vandals. A few pavers can be removed to access the well cap and then reinstated.

Long-term Inspection and Maintenance Checklist

- Inspect after at least one major storm per year. Standing water for more than 30 minutes indicates clogged openings.
- Check drain outfalls for free flow of water.
- Check outflow from observation well annually.
- Vacuum the surface openings in dry weather to remove dry, encrusted sediment at a minimum of twice annually. Sediment sometimes appear as small, dried and curled "potato chips" in the openings. Vacuum settings may require adjustment to prevent uptake of aggregates from the pavement openings and joints. Regenerative air vacuum sweepers can be used for routine removal of loose dirt from the surface and openings. Severely clogged openings can be cleaned with true vacuum equipment. Maintain vegetation around the perimeter of the pavement to filter runoff.
- Repair all ruts or deformations in the pavement exceeding 13 mm (0.5 in.).
- Repair pavers offset by more than 6 mm (0.24 in.) above/below adjacent units.
- Replace broken units that impair the structural integrity of the surface.
- Replenish aggregate void and joint materials as needed.

4.6 Cost Information

Costs vary with site activities and access, PICP depth, drainage, curbing and underdrains (if used), labor rates, contractor expertise, and competition. For vehicular applications over 1,500 m² (15,000 sf), costs generally range from $30 to $40/m² ($3 to $4/sf) for the pavers, jointing, and bedding materials when mechanically installed. Base and subbase can vary in thickness and price depending on the design. An important consideration in cost analysis is that PICP serves as a pavement and as a stormwater management BMP. Therefore, costs should be appropriately divided and allocated. In addition, some PICP designs can reduce or eliminate detention facilities, underdrains, and related water conveyance pipes, thereby enabling land to be used for buildings and related income.

Cost Case Study

An infill housing project, Autumn Trails in Moline, IL, included 3,623 m² (39,000 sf) of PICP in the streets of this 32-home development (**Figure 4-3**). According to the developer, mechanically installed PICP initial costs were lower than providing conventional impervious asphalt (75 mm [3 in.] thick with an aggregate

Table 4-2 Cost Comparisons for a PICP Project

AUTUMN TRAILS, MOLINE, ILLINOIS PAVEMENT COST COMPARISON IN M² (SF) 2006		
PICP	**Asphalt**	**Concrete**
$117.85/m² ($10.95/sf) no storm drainage	$123.77/m² ($11.50/sf) with storm drainage	$161.45/m² ($15.00/sf) with storm drainage

Figure 4-3
PICP at Autumn Trails, Moline, IL
Source: ICPI

base) or concrete pavement (200 mm [7.87 in.] thick) with inlets and storm sewers. The PICP costs in 2006 are shown in **Table 4-2** with comparisons to conventional paving with drainage.

The PICP costs include 80 mm (3.15 in.) thick pavers and 50 mm (2 in.) thick bedding of Illinois Department of Transportation (IDOT) coarse aggregate CA-16. The CA-16 was also used to fill the paver openings. The base consists of 200 mm (8 in.) of CA-7 over a subbase of 200 mm (8 in.) of CA-1. The CA-1 has perforated pipe running through both sides of its cross section to facilitate the movement of excess water to French drains. Curb costs are the same for all pavements. PICP was also selected because it was capable of winter construction.

4.7   Stormwater Benefits

Overall benefits of PICP were outlined in the introduction to this chapter. Highlights below are benefits specific to stormwater quality and quantity management goals.

Volume Reduction and Peak Rate Control

Properly designed PICP systems can provide effective management of runoff volumes and peak runoff rates from extreme storm events. A runoff volume reduction of 28% was observed for an underdrained pavement over impermeable soils (Fassman 2010), while a 100% reduction was measured for a site with no underdrains overlying hydrologic group A soils (Bean 2007). A laboratory study by Andersen (1999) simulated a 15 mm/hr (0.6 in./hr), one hour duration storm and measured that 55% of rainfall volume was retained by a completely dry pavement with a base/subbase course depth ranging from 30 to 70 cm (1.2 to 2.8 in.). Pavements that were initially wet retained 30% of the rainfall.

Pollutant Removal

Similar to other permeable pavement types, PICP systems are effective in reducing pollutants such as total suspended solids (TSS), heavy metals, motor oil, and grease. The PICP joint, bedding material, reservoir course, and the underlying soil provide pollutant removal. A study by Dierkes (2002) found that most heavy metals were captured in the top 2 cm (0.75 in.) of the aggregates in the joints. The authors concluded that through regular maintenance, where the top layer of aggregates is removed and then refilled with new material, permeable pavements can remove heavy metals over long periods of time. Another study by Gilbert (2006) observed that runoff from PICP driveways had significantly lower concentrations of TP, TN, NO_3-N, NH_3-N, TKN, TSS, Cu, Pb and Zn than runoff from asphalt driveways.

4.8   Recommendations for Specifications

PICP systems can be used for pedestrian surfaces; vehicular parking; and low-speed, low-volume roads for stormwater management. See *Interlocking Concrete Pavement Institute (ICPI) Permeable Interlocking Concrete Pavements: Selection Design Construction Maintenance.* The following specification outline must be modified and completed for project-specific conditions by a qualified design professional. This specification outline applies to full- and partial-infiltration designs.

Recommended Outline for Specifications: Permeable Interlocking Concrete Pavement (PICP) System

PART 1—GENERAL

1.01 Summary

A. **Scope of work**

This work consists of a permeable interlocking concrete pavement system for stormwater management and includes:

1. Excavation
2. Geotextile
3. Subbase of large aggregate for water storage and infiltration
4. Base of small aggregate for water storage
5. Bedding course of smaller aggregate for the concrete pavers
6. Concrete pavers with openings and/or joints
7. Joint and/or opening aggregate to fill the concrete paver openings and/or joints
8. Associated drainage inlets, outlets, monitoring well(s) and piping

B. **Related sections**

1. Section to be filled by Engineer. Concrete curbs

1.02 Submittals

A. **Product data:** Submit manufacturer's descriptive data for geotextile. Submit concrete paver manufacturer's descriptive data for the concrete pavers.

B. **Materials**

1. Four (4) representative full-size samples of each concrete paver type, thickness, color and finish. Submit samples indicating the range of color in the finished installation.
2. Laboratory test reports indicating compliance of the concrete pavers with ASTM C936 including a minimum average compressive strength of 55 MPa (8,000 psi) and average absorption no greater than 5% when tested according ASTM C140 or CSA A231.1.
3. Minimum 1.4 kg (3 lb) samples of aggregates used in the subbase, base, bedding course, and in the concrete paver openings and/or joints.
4. Sieve analysis per AASHTO T-27 indicating compliance with specified gradations of the aggregates used in the subbase, base, bedding course, and in the concrete paver openings and/or joints.

C. **Certificates from PICP Installer:** Submit the following with the bid documents:

1. Verification of current Interlocking Concrete Pavement Institute (ICPI) certificate requirements

2. The PICP installation crew shall employ at least one person holding a current ICPI PICP Installer Specialist Course record of completion who must be on site to function as project foreman, overseeing each installation crew during all PICP installations.

D. **Certificates from project inspector(s):** PICP project inspectors shall hold a PICP Installer Specialist Course record of completion.

Note: Use for LEED projects is an option for the designer. See ICPI Tech Spec 16 *Achieving LEED Credits with Segmental Concrete Pavement* for additional information..

E. **LEED v4 Submittals**

1. Sustainable sites: Calculations to demonstrate compliance with rainwater management and/or solar reflectance.

2. Water efficiency: Design and calculations to demonstrate water savings for irrigation and/or grey water building use.

3. Materials and resources: Documentation for building product disclosure and optimization including environmental product declarations, sourcing of raw materials and/or material ingredients, and/or construction and demolition waste management.

1.03 Quality Assurance

A. **Pre-construction meeting:** Conduct pre-construction meeting to review requirements for construction and protection of the PICP system. The general contractor shall provide the facility for the pre-operation conference. Representatives from the following entities shall be present at the conference:

1. General contractor's superintendent

2. PICP subcontractor foreman

3. Concrete paving unit manufacturer's representative

4. Testing laboratory(ies) representative(s)

5. Project engineer

The contractor shall submit a list of participants to the project engineer for approval. The complete listing shall identify each participant's name, employer, title, contact information, and role in construction of PICP. Construction operations of PICP shall not begin until the specified personnel have completed the mandatory pre-construction meeting. The following items shall be discussed and determined at the pre-construction meeting:

1. Methods for keeping all materials free from sediment during storage, placement, and on completed areas

2. Methods for checking slopes, surface tolerances, and elevations

3. Concrete paving unit delivery method(s), timing, storage location(s) on the site, staging, paving start points, and direction(s)

4. Anticipated daily paving production and actual record

5. Diagrams of paving laying/layer pattern and joining layers as indicated on the plans

6. Monitoring/verifying paver dimensional tolerances in the manufacturing facility and on-site if the concrete paving units are mechanically installed and two

7. Testing intervals for sieve analyses of aggregates and for the concrete paving units

8. Method(s) for tagging and numbering concrete unit paving packages delivered to the site

9. Testing lab location, test methods, report delivery, contents and, timing

10. Engineer inspection intervals and procedures for correcting work that does not conform to specifications in this work item

B. **Mock-ups**

1. Install a 3x3 m (10x10 ft) paver area

Note: Mechanized installations may require a larger mock up area. Consult with the paver installation contractor on the size of the mock up.

2. Use this area to determine surcharge of the bedding layer, joint sizes, and lines, laying pattern, color, and texture of the job.

3. This area will be used as the standard by which the work will be judged.

4. Subject to acceptance by owner, mock-up may be retained as part of finished work.

5. If mock-up is not retained, remove and properly dispose of mock-up.

1.04 Delivery, Storage, and Handling

A. **General:** Comply with Product Requirement Section.

B. **Comply with manufacturer's ordering instructions and lead-time requirements to avoid construction delays.**

C. **Delivery:** Deliver materials in manufacturer's original, unopened, undamaged container packaging with identification tags intact on each paver bundle.

1. Coordinate delivery and paving schedule to minimize interference with normal use of buildings adjacent to paving.

2. Deliver concrete pavers to the site in steel banded, plastic banded, or plastic wrapped cubes capable of transfer by forklift or clamp lift.

3. Unload pavers at job site in such a manner that no damage occurs to the product or existing construction.

D. **Storage and protection:** Store materials in protected area such that they are kept free from mud, dirt, and other foreign materials.

1.05 Environmental Requirements

A. Do not install in rain or snow.

B. Do not install frozen bedding materials.

1.06 Maintenance

A. Extra materials: Provide [Specify area] [Specify percentage] additional material for use by owner for maintenance and repair.

B. Pavers shall be from the same production run as installed materials.

PART 2—PRODUCTS

2.01 Materials

Note: Geotextiles are used along the perimeter (sides) of the subbase materials to prevent lateral migration of soils into the subbase materials. The designer may choose to also include a geotextile in silt and clay soils to protect soil subgrade from clogging and piping of fines into the subbase. Consultation with a geotechnical engineer is recommended.

A. **Geotextile**

    1. Conforming to AASHTO M-288 *Subsurface Drainage Requirements*, Class 1 shall be used. Apparent opening size (AOS) is determined by percent of soil subgrade passing the 0.075 mm (0.03 in.) sieve.

    2. Non-woven made of polypropylene staple fibers, designed for subsurface drainage applications; dimensionally stable, resistant to UV degradation and biological and chemical conditions normally found in soils.

Note: Edit gradations in **Table 4-3** and **Table 4-4** for size grading based on project design and locally available aggregates. Recommendations for the subbase are AASHTO M-43 or ASTM No. 2, No. 3, or No. 4 stone for the subbase with less than 5% passing the 19 mm (0.75 in.) sieve. The depth of the subbase depends on the hydrologic and structural design to withstand wheel loads. The aggregate base layer above the subbase is typically AASHTO M-43 or ASTM No. 57 stone and is 10 cm (4 in.) thick as it provides a transition layer between the subbase and bedding course. For pedestrian areas, the subbase layer can be eliminated and the No. 57 stone layer can be thickened to accommodate water storage and infiltration. Edit specifications accordingly. The bedding course is typically ASTM No. 8 stone. Aggregates used for filling openings and/or joints between the concrete pavers are typically ASTM No. 8 stone. ASTM No. 89 or No. 9 stone is acceptable.

B. **Aggregates**

    1. All aggregates shall be crushed, have a minimum of 90% fractured faces, and a Los Angeles Abrasion loss of less than 40 per AASHTO T-96. All aggregates shall be washed and have no greater than 2% passing the No. 200 sieve.

2. Subbase and base aggregates shall be clean, uniformly graded, washed, and crushed stone with minimum 32% voids per AASHTO T-19. Subbase aggregates shall conform to the gradation shown in **Table 4-3**:

Table 4-3 ASTM No. 2 Aggregate Subbase

SIEVE SIZE, mm (in.)	PERCENT PASSING (%)
75 (3)	100
63 (2.5)	90–100
50 (2)	35–70
37.5 (1.5)	0–15
19 (0.75)	0–5

3. Base aggregates shall conform to the gradation shown in **Table 4-4**:

Table 4-4 ASTM No. 57 Aggregate Base

SIEVE SIZE, mm (in.)	PERCENT PASSING (%)
37.5 (1.5)	100
25(1)	95–100
12.5 (0.5)	25–60
4.75 (No. 4)	0–10
2.36 (No. 8)	0–5

4. Bedding course gradation shall conform to ASTM No. 8 stone and as shown in **Table 4-5**:

Table 4-5 ASTM No. 8 Aggregate Bedding Layer

SIEVE SIZE, mm (in.)	PERCENT PASSING (%)
12.5 (0.5)	100
9.5 (0.4)	85–100
4.75 (No. 4)	10–30
2.36 (No. 8)	0–10
1.16 (No. 16)	0–5

5. Concrete paving units will comply with ASTM C936, including a minimum average compressive strength of 55 MPa (8,000 psi), and an average absorption no greater than 5% when tested according ASTM C140. Conduct freeze-thaw durability testing only according to ASTM C1645 if concrete paving units will be exposed to freezing conditions. Supplementary cementitious materials and chemical admixtures shall comply with ASTM C936. Concrete pavers shall be 80 mm (3.15 in.) thick for vehicular applications.

6. Opening and/or jointing aggregates gradation shall conform to ASTM No. 8 stone per **Table 4-5**. Finer gradation may be used conforming to ASTM No. 89 or No. 9 stone.

Note: Add more information about these components if needed. PICP requires overflow pipes or an area to accommodate overflow conditions. A monitoring well at or near the lowest elevation is highly recommended to monitor drawdown times. This well is typically a 100 to 150 mm (4 to 6 in.) diameter perforated pipe set vertically with a screw cap.

C. **Drainage inlets, outlets and piping:** Provide as shown on the plans.

PART 3—EXECUTION

3.01 Preparation and Excavation

Note: During the project planning stages, determine if it is necessary to construct the PICP system before other soil-disturbing construction is completed to comply with project requirements. At a minimum:

1. Protect finished PICP system by covering the surface with a second geotextile and a 50 mm (2 in.) thick open-graded aggregate layer, or;

2. Establish temporary roads for site access that do not allow vehicular traffic to contaminate the PICP materials and surface with sediment, or;

3. Construct the aggregate subbase and base, and protect the surface of the open-graded base aggregate with geotextile and a 50 mm (2 in.) thick layer of the same base aggregate over the geotextile. Thicken this layer at transitions to match elevations of adjacent pavement surfaces subject to vehicular traffic. When other construction is completed, first remove geotextile, then remove soiled aggregate, and finally install the remainder of the PICP system.

A. **PICP Schedule:** Schedule work late in the project schedule after grading and landscaping are completed to reduce potential exposure to dust and silty runoff.

Note: It is essential to protect the work from runoff and debris. Both the pavement and the underlying stone courses risk clogging without protection.

B. **General:** Before placing PICP system, complete any adjacent work such as guardrails, site cleanup, and plantings that could damage the PICP surface and system. Any excess thickness of soil that has accumulated or applied over the excavated soil subgrade to trap sediment from adjacent construction activities shall be removed before application of the geotextile and subbase aggregate. Keep area where pavement is to be constructed free from sediment during entire job. Geotextiles and any aggregates contaminated with sediment shall be removed and replaced with clean materials. Do not damage drainpipes, overflow pipes, monitoring wells, or any inlets and other drainage appurtenances during installation. Report any damage immediately to the project engineer.

Note: It is important that the subgrade remain uncompacted so that water will infiltrate properly.

C. **Excavate the soil subgrade using equipment with tracks or over-sized tires.** Do not use heavy equipment or vehicles with narrow rubber tires. Do not compact the subgrade. Bring subgrade to line, grade, and elevations indicated on the drawings, providing a level bed to promote uniform infiltration.

D. **Avoid compacting the soil by protecting the excavated area from vehicular and equipment traffic.** Do not store materials or equipment in this area.

3.02 PICP System Installation

A. **Geotextile**

1. Perimeter geotextile: Geotextile shall be placed on the perimeter of the PICP system in all applications. Place geotextile on the perimeter of the aggregate subbase, base, and bedding layers to prevent lateral movement of soil into the PICP system.

2. Subgrade geotextile: If geotextile is specified by the designer for the soil subgrade at the soil subbase interface, it should be placed immediately after excavation. The geotextile on the soil subgrade should overlap at the edges by at least 300 mm (12 in.).

B. **Drain pipes:** Install drainage pipes and other drainage improvements as shown on the plans.

C. **Open-graded aggregate subbase**

1. Moisten, spread, and compact the AASHTO No. 2 subbase in 200 mm (8 in.) thick lifts without wrinkling or folding the geotextile. Place subbase to protect geotextile from wrinkling under equipment tires and tracks.

2. For each lift, make at least two passes with a minimum 10 T (10 ton) vibratory roller in the vibratory mode then at least two in the static mode until there is no visible movement of the No. 2 stone. A 60 kN (13,500 lbf) plate compactor with a compaction indicator may be used in lieu of a vibratory roller. Do not crush aggregate with the roller or the plate compactor. Do not crack or crush the drain pipes during compaction. The surface tolerance of the compacted No. 2 subbase shall be ±63 mm (2.5 in.) over a 3 m (10 ft) straightedge.

Note: After construction of the subbase, concrete curbing is typically formed and placed. Then the aggregate base, bedding, and pavers are installed.

D. **Open-graded aggregate base**

1. Moisten, spread, and compact the No. 57 stone base layer in one 100 mm (4 in.) thick lift. Make at least two passes with a minimum 10 T (10 ton) vibratory roller in the vibratory mode then at least two in the static mode until there is no visible movement of the No. 2 stone. A 60 kN (13,500 lbf) plate compactor with a compaction indicator may be used in lieu of a vibratory roller. Do not crush aggregate with the roller or the plate compactor. The surface tolerance of the compacted No. 57 base shall be 20 mm (± 0.75 in.) over a 3 m (10 ft) straightedge.

E. **Aggregate bedding course**

1. Moisten, spread, and screed the No. 8 stone bedding material to a 50 mm (2 in.) thickness. Fill voids left by removed screed rails with No. 8 stone.

2. The surface tolerance of the screeded No. 8 bedding layer shall be ±10 mm (0.4 in.) over a 3 m (10 ft) straightedge.

3. Do not subject screeded bedding material to any pedestrian or vehicular traffic before paving unit installation begins. Do not compact the bedding course.

F. Concrete pavers

1. Lay the paving units in the pattern(s) and joint widths shown on the plans. Maintain straight pattern lines.

2. Fill gaps at the edges of the paved area with cut units. Cut pavers subject to tire traffic shall be no smaller than one-third of a whole unit. Pavers shall be cut with a powered masonry saw prior to placement.

G. Aggregate for joints and/or openings

1. Fill concrete paver openings and/or joints with aggregate by sweeping. Remove excess aggregate on the surface by sweeping the concrete pavers clean prior to compacting the pavers.

2. Compact and seat the pavers into the bedding material using a low-amplitude, 75 to 90 Hz plate compactor capable of at least 22 kN (5,000 lb) centrifugal compaction force. Make at least two passes across the entire surface with the plate compactor.

3. Remove and replace any cracked concrete paving units with whole units.

4. Do not compact within 2 m (6 ft) of the unrestrained edges of the paving units.

5. Apply additional aggregate to the openings and/or joints if needed, filling them to within 6 mm (0.24 in.) of the top of the chamfers on the concrete paving units or to within 6 mm (0.24 in.) of the top surface if the concrete paving units have no chamfers. Remove excess aggregate by sweeping.

6. All concrete paving units must have openings and/or joints filled with aggregate and compacted within 2 m (6 ft) of the laying face at the completion of each day.

7. The final surface tolerance of compacted pavers shall not deviate more than ±10 mm (0.4 in.) under a 3 m (10 ft) long straightedge.

H. After work in this section is complete, the general contractor shall be responsible for protecting the concrete paver surface from sediment deposition and damage due to subsequent construction activity on the site.

I. Do not apply sand to PICP during construction or during service.

3.03 Field Quality Control

A. Surface characteristics

1. After sweeping the surface clean, check final elevations for conformance to the plans.

2. Lippage: No greater than 3 mm (0.12 in.) difference in height between adjacent pavers.

3. The surface elevation of pavers shall be 3 to 6 mm (0.12 to 0.24 in.) above adjacent drainage inlets, concrete collars, or channels.

4. Joint lines for paver courses: ±15 mm (0.6 in.) over a 15 m (50 ft) string line.

5. Test the permeability of the complete pavement surface according to ASTM C1781. Perform test at three locations for areas up to 2,500 m² (25,000 sf). Add one test location for each additional 1,000 m² (10,000 sf) or fraction thereof. Provide at least 1 m (3 ft) clear distance between test locations, unless at least 24 hours have elapsed between tests. Do not conduct the test if there is

standing water on top of the PICP. Do not conduct tests within 24 hours of the last precipitation. Average minimum surface infiltration shall be 254 cm/hr (100 in./hr). Report test results to the project engineer or owner's representative.

B. **Replace work that does not meet standards.**

C. **The PICP installation contractor shall return to site after 6 months from the completion of the work and provide the following as required:** fill paver joints with stones, replace broken, or cracked pavers and re-level settled pavers to specified elevations. Any rectification work shall be considered part of original bid price and with no additional compensation.

4.9   General Resources

Ferguson, B. (2005). "Porous Pavements." CRC Press, Boca Raton, Florida.

Interlocking Concrete Pavement Institute (ICPI) (2009). "Permeable Design Pro"(Software) for hydrologic and structural design of PICP.

Smith (2011). "Permeable Interlocking Concrete Pavements—Design, Specification, Construction, and Maintenance." Interlocking Concrete Pavement Institute, 4th Ed., Interlocking Concrete Pavement Institute (ICPI), Herndon, VA.

Websites:

Interlocking Concrete Pavement Institute (ICPI) PICP resources for design, construction and maintenance <www.icpi.org>.

Low Impact Development Center <www.lowimpactdevelopment.org>.

North Carolina State University <www.bae.ncsu.edu/info/permeable-pavement>.

4.10 References

American Association of State Highway and Transportation Officials (AASHTO) (1993). "Guide for Design of Pavement Structures." American Association of State Highway and Transportation Officials, Washington, D.C.

American Disabilities Act (ADA) (2010). 2010 ADA Standards for Accessible Design, United States Department of Justice, Washington, D.C.

American Society for Testing and Materials (ASTM) (2013). "ASTM C936 Standard Specification of Solid Concrete Interlocking Paving Units." *Annual Book of Standards*, American Society for Testing and Materials, 4(5), Conshohocken, PA.

Andersen, C. T., Foster, I. D. L., and Pratt, C. J. (1999). "Role of Urban Surfaces (Permeable Pavements) in Regulating Drainage and Evaporation: Development of a Laboratory Simulation Experiment." *Hydrological Processes,* 13(4): 597.

Bean, E. Z., Hunt, W. F., Bidelspach, D. A., and Smith, J.E., (2004). "Study on the Surface Infiltration Rate of Permeable Pavements." *Proc., 1st Water and Environment Specialty Conference of the Canadian Society for Civil Engineering*, Saskatoon, Saskatchewan, Canada.

Bean, E. Z., Hunt, W. F., and Bidelspach, D. A., (2005). "A Monitoring Field Study of Permeable Pavement Sites in North Carolina." *Proc., 8th Biennial Conference on Stormwater Research & Watershed Management.*

Bean, E. Z., Hunt, W. F., and Bidelspach, D. A. (2007). "Evaluation of Four Permeable Pavement Sites in Eastern North Carolina for Runoff Reduction and Water Quality Impacts." *Journal of Irrigation and Drainage Engineering,* 133(6), 583-592.

Booth, D. B., and Leavitt, J. (1999). "Field Evaluation of Permeable Pavement Systems for Improved Stormwater Management." *American Planning Association Journal*, 65(3), 314–325.

Brattebo, B. O., and Booth, D. B. (2003). "Long-Term Stormwater Quantity and Quality Performance of Permeable Pavement Systems." *Water Resources*, Elsevier Press.

Canadian Standards Association (CSA) (2014). *CSA A231.1 Precast Concrete Paving Slabs and CSA A231.2 Precast Concrete Pavers*, Rexdale, Ontario.

Chopra, M. B., Stuart, E., and Wanielista, M. (2010). "Pervious Pavement Systems in Florida—Research Results." *Proc., 2010 International Low Impact Development Conference*, San Francisco, CA, American Society of Civil Engineers (ASCE), Reston, VA.

Collins, K. A., et al. (2008). "Hydrologic Comparison of Four Types of Permeable Pavement and Standard Asphalt in Eastern North Carolina." *Journal of Hydrologic Engineering*.

Dierkes, C., Kuhlmann, L., Kandasamy, J., and Angelis, G. (2002). "Pollution Retention Capability and Maintenance of Permeable Pavements." *Proc., 9th International Conference on Urban Drainage, Global Solutions for Urban Drainage*, Portland, OR, American Society of Civil Engineers (ASCE), Reston, Virginia.

Fassman, E., and Blackbourn, S. (2006). "Permeable Pavement Performance for Use in Active Roadways in Auckland, New Zealand." University of Auckland, Auckland, New Zealand.

Fassman, E. and Blackbourn, S. (2010). "Urban Runoff Mitigation by a Permeable Pavement System Over Impermeable Soils." *Journal of Hydrologic Engineering*. 15 (6): 475–485.

Gerrits, C. and James, W. (2002). "Restoration of Infiltration Capacity of Permeable Pavers." *9th International Conference on Urban Drainage*, Portland, OR, American Society of Civil Engineers, Reston, VA.

Gilbert, J. K., and Clausen, J. C. (2006). "Stormwater Runoff Quality and Quantity from Asphalt, Paver, and Crushed Stone Driveways in Connecticut." *Water Research*. 40: 826–832.

Hawkins, R., Ward, T. J., Woodward, D. E., and Van Mullem, J. A. (2009). "Curve Number Hydrology: State of the Practice, Environmental and Water Resources Institute." American Society of Civil Engineers, Reston, VA.

Interlocking Concrete Pavement Institute (ICPI) (2006). "Permeable Interlocking Concrete Pavements—Design, Specification, Construction, and Maintenance" 3rd Ed., Interlocking Concrete Pavement Institute (ICPI).

Interlocking Concrete Pavement Institute (ICPI) (2009). "Permeable Design Pro" (Software) for hydrologic and structural design of PICP.

Kinter, M. (2010). "Maintenance and Restoration of Porous Pavement Surfaces." Elgin Sweeper Company, Elgin, IL.

Mallela, J., Larson, G., Wyatt, T., Hall, J. and Barker, W. (2002). "User's Guide for Drainage Requirements in Pavements—Drip 2.0 Microcomputer Program." FHWA IF 02 05C, Office of Pavement Technology, Federal Highway Administration, United States Department of Transportation, Washington, D.C.

Natural Resource Conservation Service (NRCS) (1986). "Urban Hydrology for Small Watersheds." *Technical Release 55*, United States Department of Agriculture, Natural Resource Conservation Service, Washington, D.C.

North Carolina State University and North Carolina A&T State University Cooperative Extension (2009). "Urban Waterways—Permeable Pavement: Research Update and Design Implications." E08-50327.

Sustainable Technologies Evaluation Program (STEP) (2009). "A Review of Permeable Pavement Research from the University of Guelph." Toronto and Region Conservation Authority, Downsview, Ontario, Canada.

Van Seters, T. (2007). "Performance Evaluation of Permeable Pavement and a Bioretention Swale." Seneca College, King City, Ontario, Interim Report #3, Toronto and Region Conservation Authority, Downsview, Ontario.

Grid Pavement Fact Sheet

DESCRIPTION

Grid pavements are comprised of concrete or plastic open-celled paving units. The "cells" or openings penetrate their entire thickness so they can accommodate aggregate, topsoil, or grass. Concrete and plastic grids are intended for light vehicular loading applications and are typically constructed over a dense-graded aggregate base. Both types of grids are often used for emergency access drives and parking/drive lanes with occasional use, where a natural turf appearance and infiltration are desired as well as where high intensity uses or loads are not expected. In some cases, open-graded aggregate within the grid openings and an open-graded base are used with these products for additional stormwater storage and infiltration.

Source: Hastings Pavement Company

APPLICATIONS

POTENTIAL APPLICATION		NOTES
Overflow Parking	Yes	Primary use
Primary Parking Areas (most heavily used)	Limited	Grass can die under engine heat and wheel tracking; aggregate infill required in parking areas subject to constant use; base thickness must consider design loads and use of parking lot
Sidewalks/Pathways	Yes	Recreational paths and non-street walks
Drive/Aisles	Yes	Base thickness must consider design loads and use of parking lot; can be used for boat ramps
Roads/Highways	Limited	Can be used for median crossovers, shoulder, maintenance access roads, parking, rest areas, etc.; base thickness must consider design loads and use of parking lot
Access Drives/Ring Roads	Yes	Base thickness must consider design loads and use of parking lot
Loading Areas	No	
Frequent Truck Traffic	Limited	Parking areas only; not designed for truck traffic areas or areas of repeated turning movements; base thickness must consider design loads and use of parking lot

Source: © ICPI

Grid Pavement Fact Sheet continued

BENEFITS

- Turf or gravel pavement surface capable of limited vehicular traffic that reduces runoff and provides a cooler surface than conventional pavements.
- Openings in grids can be filled with soil and grass or washed and highly permeable aggregates.
- Provides moderate volume and peak rate reduction, which can be increased with open-graded base storage
- Provides moderate groundwater recharge, which can be increased with open-graded base storage

STORMWATER QUANTITY FUNCTIONS		NOTES
Volume Reduction	Low/Medium	Volume control increased with additional storage provided in the aggregate base/subbase
Groundwater Recharge	Low/Medium	Recharge high if soils permit
Peak Flow Rate Reduction	Low/Medium	Peak discharge rate control increased with increased storage and/or high soil permeability
STORMWATER QUALITY FUNCTIONS		NOTES
Total Suspended Solids	High	Pre-treatment encouraged but not required
Total Nitrogen/ Total Phosphorus	Low/ Moderate	Depends on design; sand in grid openings can help reduce nitrogen
Metals	High	
Temperature	High	Maintains and does not increase ambient rainfall temperatures; grass provides microclimatic cooling, filtering in subbases provides additional cooling

SITE CONSTRAINTS/CONCERNS

- Not recommended for high intensity traffic uses or heavy loads
- If grassed, requires additional time before using for grass to establish and may require reseeding, fertilization, and initial irrigation; salt and heavy use may kill grass.
- Limit or prohibit runoff from contributing impervious surfaces to minimize sediment deposition
- Feasibility of steeper slopes (<5%) are possible with design modifications such as berms or barriers in the subgrade to control lateral flows and promote infiltration.

Grid Pavement Fact Sheet continued

RECOMMENDED KEY DESIGN CRITERIA

- Use minimum 200 mm (8 in.) thick compacted, dense-graded aggregate base layer to support vehicular loads. Thicker bases/subbases may be required over weak or continually wet soils. Consult specifically with geotechnical or pavement engineer.
- Concrete grids: Typical 80 mm (3.15 in.) thick units conforming to ASTM C1319
- Provide stationary edge restraints at perimeter to constrain grid paving units

PERFORMANCE

LOW/MEDIUM—Intended for low vehicular use areas; can withstand occasional truck traffic including fire trucks. Larger openings may result in less clogging.

OPERATION & MAINTENANCE

MEDIUM—Grass requires maintenance such as mowing, fertilizing, re-seeding, and watering; maintenance reduced with use of aggregates in openings.

COSTS

MEDIUM—For minimum 1,500 m² (15,000 sf) area using prevailing labor wages: $30 to $40/m² ($3 to $4/sf) excluding base and drainage.

REFERENCES

Interlocking Concrete Pavement Institute (ICPI) (1999). *ICPI Tech Spec 8 Concrete Grid Pavements*, Interlocking Concrete Pavement Institute, Herndon, VA, <www.icpi.org>.

Ferguson, B. (2005). *Porous Pavements*, CRC Press, Boca Raton, FL.

Websites:

Interlocking Concrete Pavement Institute (ICPI) PICP resources for design, construction and maintenance <www.icpi.org>.

Invisible Structures <www.invisiblestructures.com>.

Low-Impact Development Center <www.lowimpactdevelopment.org>.

North Carolina State University <www.bae.ncsu.edu/info/permeable-pavement>.

5 Grid Pavement

5.1 System Description

Open grid pavements consist of concrete or plastic units with large surface openings filled with a permeable joint material, typically small aggregate (ASTM No. 8 or No. 89), sand, or topsoil and grass. Plastic grid units are placed directly over an open-graded aggregate base or over a thin bedding sand layer on a dense-graded base. Concrete grid units are typically placed over a thin sand bedding layer and dense-graded aggregate layer. Geotextile may be used in the design to provide separation of the aggregate base from the soil subgrade. As described in Chapter 1, the open-graded aggregate in the openings, bedding, and base provide water storage and infiltration.

Concrete Grid Paving Units

Concrete grid units are defined in ASTM C1319 *Standard Specification for Concrete Grid Paving Units*, as having maximum dimensions of 610 mm long by 610 mm wide (24 in. x 24 in.) and a minimum nominal thickness of 80 mm (3.15 in.). The minimum required thickness of the webs between the openings is 25 mm (1 in.). Dimensional tolerances should not differ from approved samples by more than 3.2 mm (0.13 in.) for length, width, and height. Concrete grid paving units may include steel wire or fiber reinforcement to increase flexural strength.

The minimum compressive strength of the concrete grid units should average 35 MPa (5,000 psi) with no individual unit less than 31 MPa (4,500 psi) per ASTM C140 *Standard Test Methods for Sampling and Testing Concrete Masonry Units and Related Units*. Average water absorption among tested samples should not exceed 160 kg/m³ (10 lb/ft³). Freeze-thaw durability is based on three years of proven field performance of units that meet the web thickness, compressive strength, and absorption criteria as previously described.

Concrete grid unit designs fall into two categories: lattice and castellated (**Figure 5-1**). Lattice pavers have a flat surface that forms a continuous pattern of concrete when installed. Castellated grids have protruding concrete

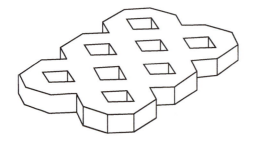 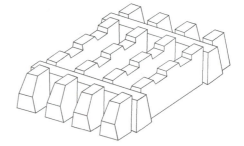

Figure 5-1
Examples of lattice and castellated concrete grids
Source: ICPI 2014

knobs on the surface. If seeded, this design makes the grass appear continuous when installed. Concrete grid pavers generally range from 20 kg (45 lb) to 40 kg (90 lb). The open surface area is between 20% and 70%.

Plastic Grid Paving Units

Figure 5-2 depicts plastic grid applications using aggregate (gravel) in the surface with an open-graded base (ASTM No. 2 stone) for water storage and infiltration. Similar plastic grid paving units, or "geocells," are sold under various product names. While some plastic geocells are designed specifically for base or soil stabilization applications and not necessarily permeable pavement, they may be used as a surfacing that stabilizes aggregate or topsoil and grass under light vehicular traffic. Designers should confirm surface pavement uses with plastic grid geocell manufacturers.

Figure 5-2
Plastic grid using aggregate in the openings
Source: Geosyntec Consultants, Inc.

Plastic grids can be made of recycled materials. At the time of this publication, no governing ASTM or other product standard for testing or industry consensus guidelines for installation were defined. However, plastic grids are typically placed on, or anchored to, a compacted aggregate base.

Plastic and concrete grid pavements are intended for infrequently used parking areas and travel lanes. At this time, they are not recommended for loading areas or in commercial parking lots that receive high daily use. Grid pavers are suitable replacements for conventional asphalt or concrete pavements in light duty areas that provide a visually pleasing, durable surface that resists erosion and rutting, typical to non-reinforced grass or gravel pavements.

5.2 Applications—Site Constraints/Concerns

Applications and General Site Selection Criteria

- **Uses**—Walks, parking lots, light duty drives around commercial, institutional, recreational, and cultural buildings
- **Slopes**—Typical surface slopes are less than 5%.
- **Run-on**—No post-construction increase in impervious cover draining into the pavement unless the pavement is designed to infiltrate and store runoff from future increases in impervious cover; Any increased run-on must be free from sediment or organic matter.
- **Grass filled grids**—Grid pavements filled with topsoil and grass are best suited for applications such as fire lanes, maintenance access roads, and temporary event or overflow parking. Pavers using grass should be used in low or light traffic areas with no use of deicers so grass growth is not inhibited. Grass and gravel filled grid pavements are suitable for vehicular applications where lifetime loads do not exceed 7,500 80 kN (18,000 lb) single axle loads or ESALs. This accommodates a limited number of loads from trucks and fire trucks, making them suitable for occasional truck access and emergency use.
- **Aggregate filled grids**—Available for immediate use following installation; cannot use grass until well established.

Climate Considerations

- Aggregate-filled grids are more suited for areas where grass will not survive well (e.g., arid climates) or cannot be maintained.

- Snow can be plowed from plastic or concrete grids if the plow blade is set slightly above their surface so that grass is not removed. Rotary brushes for snow removal are not recommended. Deicing salts should never be used on grass because salt will kill it. Re-establishing grass in openings with contaminated soil is difficult without removing and replacing the soil.

5.3   Design Considerations

General Design Considerations

If intended to achieve significant stormwater infiltration, grid pavement systems should follow the general design guidelines included in Chapter 1 and the following:

Specific Design Considerations

Clogging

Like other permeable pavements, grids can be susceptible to clogging with sediment. Deposited sediment can kill grass. The dead grass and sediment will require removal or possible replenishing of the topsoil and grass re-seeding. Sediment deposition can also reduce surface infiltration through aggregate surfaces and provide a medium for weed growth. If grass in grid pavements cannot be maintained by the project owner/tenant or the climate is arid and without site irrigation, then crushed stone aggregate should be placed in the openings. Aggregate also should be used if sediment from the site or adjacent areas might wash onto the grids or be deposited on them by vehicles. As a worst case situation, aggregate in the grid openings clogged with sediment, preventing any infiltration, can be removed by vacuuming and replaced with fresh aggregate.

Cracking

Due to their slab-like shape, concrete grids may crack while in service. Smaller grid units (under 300 mm [12 in.]) length and width) can reduce the risk of cracking. Reinforced grids or smaller sized units are recommended when using ASTM No. 8 stone or similar as the bedding and void fill material, as this material can increase the risk of larger units shifting and displacing under wheel loads. In most situations, one or two cracks in a unit will not diminish structural or functional performance. If units crack from soil or base settlement, they can be removed and replaced after base or subgrade repairs. Likewise, plastic and concrete grid units can be reinstated after repairs to the base or underground utilities.

Structural Design Considerations

The plastic and concrete grids themselves provide little to no structural contribution to the pavement system. However, structural contributions are provided from the aggregate base materials under them. Therefore, AASHTO layer coefficients typical for dense or reduced for open-graded bases can be used in structural design. This conservative approach recognizes that the grids serve primarily as a protective surfacing for the base and greater stability for grass or aggregate in the cells, while providing infiltration and little structural contribution.

Hydrologic Design

The hydrologic design of permeable grid pavement systems is similar to that of other permeable pavements (see **Chapter 1 and Chapter 9**). Additional considerations for grid pavements are described below.

A common error in the design of permeable pavement is assuming that the amount or percent of open surface area is equal to the percent of perviousness. For example, a 30% open surface area in a grid pavement is incorrectly assumed to be 30% pervious or 70% impervious. The perviousness and amount of infiltration are dependent on the infiltration rates of the joint filling material, bedding layer, and base materials, not the percentage of surface open area.

Modeling runoff from grid pavements is based on the infill material selected (turf or aggregate). Grid pavements with grass will have runoff coefficients between 0.25 and 0.35 with the lower figure relating to applications on sandy soils and the higher on clay (Smith 1984). The North Carolina Department of Environment and Natural Resources Stormwater BMP manual suggests a runoff coefficient of 0.22 for grids with managed turf on soils, having an infiltration rate of at least 13 mm/hr (0.52 in./hr). Similar runoff coefficients are likely for grids filled with highly permeable aggregates in the openings and are supported by a dense-graded aggregate base. Grid pavements with open-graded aggregates in the openings' bedding and base will have runoff coefficients similar to that from permeable interlocking concrete pavements. Runoff coefficients will be a function of the soil infiltration rate, water storage, and resulting outflows from the base. The coefficient of NRCS runoff curve numbers for grid pavements will vary with the design and soil subgrade, ranging from 45 to 80.

Infiltration Characteristics

A few researchers have measured infiltration rates of concrete grid pavements of varying ages and the rates reflect those similar to grass. Bean (2007) reported surface infiltration rates on 15 concrete grid pavement sites using grass or soil and clogging in North Carolina, Delaware, and Maryland. Surface infiltration rates ranged between 1 to 19 cm/hr (0.375 to 7.5 in./hr) with 6.9 cm/hr (2.75 in./hr) as an average among all sites. All sites were reported as built over open-graded aggregate bases.

Collins (2007) measured surface infiltration rates between 91 and 101 cm/hr (35.8 and 39.8 in./hr) on a new concrete grid pavement in Kinston, NC with sand (no vegetation) in the openings over an open-graded aggregate base. Collins also confirmed that the new concrete grids reduced surface runoff an average of 98% from 40 storms. Peak flows measured from an underdrain were reduced an average of 77% for 36 storms compared to surface runoff from the asphalt. Average total suspended solids were lowest from the concrete grids compared to asphalt, pervious concrete, and two permeable interlocking concrete pavement installations evaluated on the same site.

Base Course

When used as a reservoir course, the base should be composed of clean, open-graded, and compacted aggregate (often ASTM No. 57, No. 2 or No. 3 gradations) with approximately 40% void space. Dense-graded bases and the soil subgrade under them should be uniformly compacted to a minimum 95% standard Proctor density per ASTM D698.

A thicker dense- or open-graded base may be required when the soil subgrade is weak (California Bearing Ratio <4% or R-value <9), has high amounts of clay or silt, or subject to frost heaving. Likewise, thicker bases, or those stabilized with cement, may be required over a seasonally high water table that creates an unstable soil subgrade in low-lying areas subject to flooding or over continually saturated soils. For unstabilized bases, geotextile is recommended to separate the compacted soil subgrade from the compacted base material. Perforated drain pipes placed at the top of the soil subgrade can provide additional stability by helping to avoid saturated conditions that might lead to premature rutting. Grid manufacturers' literature may have further recommendations on base thickness and materials as a function of the soil strength and anticipated vehicular loads.

Compacted, crushed stone, and dense-graded bases—similar to that used under conventional asphalt paving—are acceptable under grid pavements. However, they cannot offer water storage associated with open-graded bases. Generally, dense-graded bases will store water similar to that of a gravelly soil and are not considered a permeable pavement system.

If no local standards for base materials exist, gradation of dense-graded base should conform to ASTM D2940 *Standard Specification for Graded Aggregate Material for Bases or Subbases for Highways or Airports*. A minimum of 200 mm (8 in.) thick compacted aggregate base is recommended for intermittently used parking lots, driveways, and emergency fire lanes supporting fire trucks, including occasional truck axle loads defined by AASHTO H20 and HS20. Dense-graded bases can be exposed to high levels of moisture in grid pavements. This eventuality should be considered when deciding materials, thickness, and compaction requirements for vehicular applications. At a minimum, bases and the soil subgrade under them should be uniformly compacted to a minimum 95% standard Proctor density per ASTM D698 or AASHTO T-99. This minimizes the deformation of perforated drain pipes at the bottom of the base (top of the soil subgrade) from pore pressures and wheel loads when the base material is saturated.

Concrete Grid Pavement

Concrete grids on a compacted aggregate base require a 13 to 25 mm (0.5 to 1 in.) thick layer of bedding sand under the grids. The gradation of the bedding sand should conform to ASTM C33. Gradations are given in the concrete grid pavement guide specification at the end of this Chapter 5. Limestone screenings, stone dust, or masonry sand should not be used. The sand is placed, screeded to a consistent thickness, and used over dense-graded bases only. ASTM No. 8 stone used for bedding and openings can be used over an open-grade base. The bedding layer should be screeded to 25 mm (1 in.) thickness.

Edge restraints are required for preventing concrete grid pavements from shifting under tire traffic. Concrete, plastic, or metal edge restraints are recommended where automobile tires could loosen and damage the edge units. ICPI Tech Spec 3 Edge Restraints for Interlocking Concrete Pavements provides further guidance on their selection (ICPI 1999).

The base under concrete grids should extend beyond their perimeter to a minimum of 300 mm (12 in.) when there is no building or concrete curb to restrain them. If no concrete curbs are used, the extended perimeter increases the stability of the grids and facilitates installation of edge restraints staked into the compacted aggregate base.

Vegetation Selection

The choice of grass variety is important to longevity under tires and drought. A limited amount of research on concrete grid pavers by Sherman (1980) has shown that grasses within grids can recuperate from tire traffic damage faster than without grids. The study showed that Merion Kentucky bluegrass, Kentucky 31 tall fescue, and Manhattan perennial ryegrass have a high tolerance to wear, high potential for recuperation from damage, and a low tendency toward thatch build-up. Turf specialists may have further recommendations on species, seeding rates, and maintenance.

5.4 Recommended Installation Guidelines

Responsibility for excavation; soil subgrade grading and compaction; and base installation and compaction will vary with each project. Some or all of these operations may be executed by the grid pavement installer. No matter how construction responsibilities are divided, the ability of the soil subgrade to infiltrate water should be confirmed by a geotechnical engineer. The project engineer soil should certify that elevations and slopes conform to those on the drawings. In addition, density tests from compacted soils and base materials should be approved by the project engineer. When the subgrade meets the requirements of the drawings and specifications, the installation of the aggregate should begin immediately to protect the subgrade from rainfall, sedimentation, and construction vehicles. The specifications should require that the grid pavement contractor notify the project engineer if the subgrade (or base if done by others) does not appear to meet specifications or if other problems exist.

Concrete Grid Pavement Installation

The Interlocking Concrete Pavement Institute (ICPI) offers an advanced educational program for contractors called the ICPI Commercial Paver Technician course, which includes instructional installation of concrete grid pavements. Upon successful course completion, participants earn a record of completion that should be included in the project specifications for verification. Construction specifications should also state that the installation crew employs a job foreman who remains on the jobsite and holds an ICPI record of completion. For all grid pavement projects, specifications should require project/work history at the time of bid document submission by the installation subcontractor/general contractor. A pre-construction meeting is essential to the success of all permeable pavement projects. Chapter 4 provides a list of items for discussion at a pre-construction meeting.

Once a dense-graded base is installed and compacted, a 13 to 25 mm (0.5 to 1 in.) thick layer of bedding sand can be placed. The gradation of the bedding sand should conform to ASTM C33. Gradations are given in the concrete grid pavement Section 5.8 Recommendations for Specifications at the end of this chapter. The sand should be placed and screeded to a consistent thickness. If No. 8 stone is used for bedding and openings, this should be screeded to 25 mm (1 in.) thickness. The sand should have consistent moisture content, but not be saturated or completely dry. It should not be disturbed prior to placing the grids.

Place the grids on the screeded bedding sand with the maximum joint spacing of 5 mm (0.2 in.). Grids can be placed mechanically or by hand. The units should not be pushed or hammered such that they touch each other. If the grids touch, they may crack, chip, or spall under repeated traffic. The concrete grid units should be saw-cut to fill any spaces along the edges prior to compaction. All installed units should be compacted into the bedding sand at the end of each day. Rainfall settles uncompacted sand, preventing the grids from seating into the sand when compacted. If bedding sand is left uncompacted, it should be covered with plastic to protect it from rain. Otherwise, bedding sand saturated with rainfall prior to compaction will need to be dried, raked, and re-screeded or replaced. If left uncorrected, the grids will settle unevenly and move under traffic.

After the grids are placed, the permeable fill materials are spread across them and swept into the openings. If using topsoil, fertilizer may be mixed with the topsoil. Fertilizer quantities should account for the concrete surface. The grids are vibrated into the sand with a high frequency (75-90 Hz), low-amplitude plate compactor. It should have a minimum centrifugal compaction force of 22 kN (5,000 lbs). Rollers or a mat should be attached to the plate of the compactor to protect the grids from cracking and chipping. The primary purpose of compaction is to create a level surface among the units. An occasional cracked unit from compaction will not compromise performance. Following compaction, additional joint material should be added as needed so that the grid openings are completely filled with joint material. If using topsoil, a thick layer of topsoil can be added to the entire surface to assist in germination and create a completely grassed surface. Straw can be applied to protect the grass while it is growing.

Plastic Grid Pavement Installation

Plastic grid pavements require installation on compacted aggregate base. The grids are typically placed by hand. Depending on the manufacturer's recommendations, they can be filled with stone or washed sand, and topped with sod or topsoil and seed. Other plastic grids, typically with thin walls within their cells, allow sod to be pressed directly into the cell openings.

Like concrete grids, aligned placement of the first course of plastic grids is critical to provide proper alignment with the remaining units. Plastic units can be shaped to the installation areas by cutting them. Some plastic grids require anchoring with galvanized nails that may include galvanized washers. Other plastic grids may use U-shaped steel pins. Manufacturer's instructions provide staking materials and installation patterns. Some plastic grids are filled with free-draining sand as a growing medium for seeded topsoil or sod. If used, the sand material should be installed without moving the pavers. Booth (1999) suggests that sand as a growing medium makes the grass more susceptible to drought, and topsoil within the plastic grids might be a more effective growing medium. However, Ferguson (2005) suggests that sand can resist compaction and drain more readily than fine-textured soils in the openings. Whether sand or soil, Ferguson notes that the material in the openings need to settle into the grids via compaction or by watering. Geotextile is not used in these upper layers so roots can develop and grow around the grids and into the base.

Grass Establishment

Sediment from runoff and dust from adjacent areas must be kept from entering the openings during and after grass establishment. Sediment clogs the topsoil and prevents grass from growing. The grass should not be exposed to tires until it is well established. A period of time, typically three to four weeks, for establishing grass should be part of the construction contract and schedule.

Sod and seeded topsoil must be kept moist and fertilized to achieve a thick root mat of turf. Seeded areas that do not germinate should be reseeded immediately. Establishment will normally require six to eight weeks for seeding and three to four weeks for sod. In both cases, sufficient time must be allowed for the grass roots to grow through the pavers and help anchor the pavers to the base.

Tire Stops

Tire stops are recommended in parking lots to help prevent lateral movement of perimeter units from errant vehicles and must be appropriately anchored.

5.5   Post-construction Operation and Maintenance

A maintenance check list to maintain surface infiltration of grid pavements follows:

- Biweekly inspections recommended for the first growing season or until healthy vegetation is established
- Water grass during dry weather and establishment
- Mow grass when height exceeds 75 mm (3 in.)
- Fertilize in the spring and fall (limit use if nutrients runoff to nearby water)
- Inspect surface annually for soil/grass/aggregate erosion, scour, and unwanted growth; remove sediment and stabilize source(s)/re-seed bare soil areas as needed/replenish lost aggregate

5.6   Cost Information

Costs vary with site activities and access; base type and depth; drainage; curbing and underdrains (if used); labor rates; contractor expertise; and competition. For vehicular applications over 1,500 m² (16,146 sf), costs generally range from \$30 to \$40/m² (\$3 to \$4/sf) for the paving units, topsoil, grass seed, and installation. Mechanized installation of concrete grid pavements can reduce installation costs and time. Excavation, base materials, and installation costs are excluded.

Ferguson (2005) cites a 1998 cost comparison between conventional asphalt and plastic grid pavements with grass or gravel used in an Alabama parking lot. Annual maintenance costs were about 40% less for the plastic grid system than the asphalt. The cost comparisons were a mix of hypothetical and actual costs from the facility manager. Not unexpectedly, gravel-filled plastic grids had lower maintenance costs than grass filled units.

A study by Smith (1981) for the City of Dayton, OH, identified higher initial costs for a 3,000 m² (32,292 sf) concrete grid pavement parking lot than conventional asphalt, but lower life-cycle costs for the grid pavement. Maintenance costs over 20 years on the grid pavement lot were estimated on a 0.0929 m² (1 sf) basis and included the following:

- Mowing ten (10) times annually with a 1. 4 m (56 in.) wide riding mower
- Fertilizing annually
- No regular watering
- Reseeding bare spots and watering

Maintenance on the asphalt lot over 20 years reviewed two scenarios: the first with three seal coats, parking stall restriping and a 40 mm (1.5 in.) asphalt overlay, and the second with two overlays and two seal coats with restriping. Combined construction and maintenance costs indicated that the two asphalt parking lot scenarios were respectively about 10% and 30% higher than construction and maintenance costs for a concrete grid pavement lot.

5.7   Stormwater Benefits

Reduced Runoff

- Substantial runoff reduction comparable to grass-only surfaces
- Up to 100% infiltration for common storms consisting of short duration and low rainfall intensity

Improved Water Quality

- Reduces nutrients, metals and oils
- Does not raise runoff temperature which can damage aquatic life
- Reduced surface temperature compared to asphalt

Site Utilization

- Enables overflow parking and emergency access lanes to be grassed, permeable surfaces that blend into adjacent vegetated areas
- Promotes tree survival by providing air and water to tree roots

Drainage System

- Reduced downstream flows and possible stream bank erosion due to decreased peak flow rates and volumes
- Increased recharge of groundwater
- Reduced peak discharges to storm drains
- Reduced discharge to combined sanitary/sewer systems if present

Reduced Operating Costs

- Reduced overall project costs due to reducing or eliminating piped storm drain systems
- May enable landowner credits towards stormwater utility fees

Supporting Research

Since the early 1980s, water quality and pollutant reduction studies have been done by several researchers. In 1980, Day built a laboratory apparatus consisting of three grid pavements and conventional concrete pavement, each 1.8 m^2 (19.4 sf)(Day 1980). The grid openings were filled with topsoil and grass, placed over a 25 to 50 mm (1 to 2 in.) thick layer of sand and 150 mm (6 in.) thick ASTM No. 57 stone bases. The bases were and placed over a 250 to 300 mm (10 to 12 in.) thick layer of compacted local clay soils (with varying infiltration rates) with outflow drainage. An irrigation nozzle was used as a rain simulator. The research identified a range of runoff coefficients for concrete grid pavements since none existed in product the literature at the time. He tested the grids and conventional pavement at 2%, 4%, and 7% slopes under one hour of rainfall and recorded runoff coefficients between 0 and 0.56.

For a second experiment a year later, Day (1981) used a more refined rain simulator and dosed its outflow with ten typical urban runoff pollutants while simulating ten rain events. Using the same cross section and one clay soil type for the subgrade, he found that the conventional impervious concrete pavement generated runoff coefficients from 0.58 to 0.92, while the grids had zero or almost no runoff coefficient. The grids greatly reduced pollutants, especially metals. Nitrate/nitrite concentrations increased due to leaching from the soil.

Smith (1981) reports surface runoff coefficients between 0 and 0.35 from eleven rain events from the previously mentioned concrete grid parking lot in Dayton, OH. In addition, dry bulb and radiometric temperatures were measured on this lot and an adjacent asphalt parking lot. Air temperatures were about one degree centigrade lower and radiometric temperatures were several degrees centigrade lower on the concrete grid parking lot.

Under a US EPA grant, Goforth (1984) developed a hydrologic model for porous pavements based on monitoring porous asphalt, concrete grid pavement, a gravel trench plus conventional asphalt, and concrete pavements in Austin, Texas. The experiment included monitoring of a ~600 m² (~6,458 sf) concrete grid pavement with Bermuda grass over a sand bedding layer placed directly on a sandy loam soil with no base aggregate materials to support the grids. Runoff coefficients from three short (sprinkler) simulated rain events (11 to 12 minutes) ranged between 0.18 and 0.23 and water storage from 1.63 to 2.08 cm (0.64 to 0.82 in.) with the grids and sand. Flow-weighted average concentrations from the three rain events were lowest from the concrete grid lot for total suspended solids (TSS), total nitrogen, and lead compared to the other four surfaces. A range of semi-volatile compounds were monitored from the porous asphalt underflow and concrete grid lot surface runoff. In some instances, values were lower for the concrete grid lot. This lot did not have the advantage of an open-graded base that might have reduced such pollutants further.

Brattleboro (2003) monitored a parking lot designed with three permeable pavement types including plastic (grass and gravel), concrete grids with grass, and permeable interlocking concrete pavement. He compared infiltrated water to runoff from an impervious asphalt surface. The Renton, WA site was subject to many (typical) low-intensity rainstorms. Practically no runoff was measured from any of the permeable pavements from fifteen rain events monitored. After six years of use, minor distresses in the plastic grids from parked cars were reported, but no distresses were reported for the concrete grids or permeable interlocking concrete pavement. For water quality, none of the permeable surfaces had detectable oil or lead in water sampled from collection troughs placed 300 mm (1 ft) below the pavement surface. Water hardness and conductivity were higher from the concrete pavements than those with plastic grids. Copper and zinc were lower in water collected under the all three permeable pavements compared to that in the asphalt runoff.

Collins (2007) monitored a municipal parking lot in Kinston, NC, consisting of pervious concrete, two types of permeable interlocking concrete pavement, and concrete grid pavement filled with sand and asphalt pavement. The site was located in poorly drained soils and all permeable sections were underlain by similar thicknesses of open-graded aggregate base with perforated pipe underdrains. All sections were monitored from June 2006 to July 2007 for hydrologic and water quality.

Not surprisingly, all permeable pavements substantially reduced surface runoff volumes and peak flow rates when compared to standard asphalt. The grids demonstrated better nutrient removal due to the sand in the openings (rather than grass or ASTM No. 8 stone). The conclusions noted that various permeable pavement types be treated similarly with respect to runoff reduction (assuming similar base/subbases). The report recommended further research on nitrogen or phosphorus removal so that the permeable pavement systems can earn credits in North Carolina stormwater regulations.

5.8   Recommendations for Specifications

The following information are guidelines for preparing specifications for concrete and plastic grid pavements. These are not meant to be complete and must be approved by the engineering team. Specifications should be updated as technology advances and revised for specific site use and conditions.

Concrete Grid Pavement

Note: This guide specification is for concrete grid units placed on a sand bedding course over a compacted dense-graded aggregate base. The specification allows an option of topsoil and grass in the grid openings over bedding sand or No. 8 open-graded aggregate in the grid openings and for the bedding course. This specification is for limited vehicular applications such as access roads, emergency fire lanes, and intermittently used overflow parking areas. Specifications must be edited to suit specific requirements for each project. Edit as necessary to identify the design professional in the general conditions of the contract.

If the area is exposed to recurring vehicular traffic and additional stormwater storage in the base is desired, the specifier should consider using permeable interlocking concrete pavements (PICP). PICP provide additional structural support to vehicles and runoff storage in an open-graded, crushed stone base. In such cases, the specifier should refer to Chapter 4 of this guide.

Recommended Outline for Specifications: Concrete Grid Pavement

PART 1—GENERAL

1.01 Summary

A. Section

1. Concrete grid units

2. Bedding sand

3. Edge restraints

4. Geotextiles

5. Topsoil and grass for the grid openings

6. Open-graded aggregate for the grid openings

7. Open-graded aggregate bedding course

B. Related sections

1. Section [____]: Curbs and drains

2. Section [____]: Dense-graded aggregate base

3. Section [____]: Open-graded aggregate base

1.02 References

A. American Society of Testing Materials (ASTM)

1. C33 *Specification for Concrete Aggregates*

2. C136 *Method for Sieve Analysis for Fine and Coarse Aggregate*

3. C140 *Standard Test Methods of Sampling and Testing Concrete Masonry Units*

4. C979 *Standard Specification for Pigments for Integrally Colored Concrete*

5. C1319 *Standard Specification for Concrete Grid Paving Units*

6. ASTM D698 *Standard Test Method for Laboratory Compaction Characteristics of Soil Using Standard Effort* 600 kN-m/m^3 (12,000 ft-lbf/ft^3)

7. D2940 *Standard Specification for Graded Aggregate Material for Bases or Subbases for Highways or Airports*

8. D5268 *Specification for Topsoil Used for Landscaping Purposes*

B. Interlocking Concrete Pavement Institute (ICPI)

1. ICPI Tech Spec 8—Concrete Grid Pavements

1.03   Submittals

A. **In accordance with conditions of the contract and Division 1 Submittal Procedures Section**

B. **Manufacturer's drawings and details:** Indicate perimeter conditions, relationship to adjoining materials and assemblies, expansion and control joints, paving grid, layout, patterns, color arrangement, installation, and setting details.

C. **Sieve analysis per ASTM C136 of bedding and base materials**

Note: Include D (below) if the grid openings will be filled with topsoil and grass seed, or sod plugs.

D. **Source and content of topsoil and grass seed (sod)**

E. **Concrete grid units**

   1. Color selected by architect

   2. Four (4) representative, full-size samples of each grid type, thickness, color, and finish that indicate the range of color variation and texture expected in the finished installation

   3. Accepted samples become the standard of acceptance for the work

   4. Test results from an independent testing laboratory for compliance of grid paving unit requirements to ASTM C1319

   5. Manufacturer's certification of concrete grid units by ICPI as having met applicable ASTM standards

   6. Manufacturer's catalog literature, installation instructions, and material safety data sheets for the safe handling of the specified materials and products

1.04   Quality Assurance

A. **Paving subcontractor qualifications**

   1. Engage an experienced installer who has successfully completed grid pavement installations similar in design, material, and extent indicated for this project

   2. Holds a record of completion in Interlocking Concrete Pavement Institute Commercial Paver Technician Course

B. **Single-source responsibility:** Obtain each color, type, and variety of grids, joint materials, and setting materials from single sources with resources to provide products and materials of consistent quality, appearance, and physical properties without delaying progress of the work.

C. **Regulatory requirements and approvals:** Specify applicable licensing, bonding, or other requirements of regulatory agencies.

D. **Mock-Up**

   1. Locate where directed by the architect

   2. Notify architect in advance of dates when mock-ups will be erected

3. Install minimum 10 m² (108 sf) of concrete grid units

4. Use this area to determine the quality of workmanship to be produced in the final unit of work including surcharge of the bedding sand layer, joint sizes, lines, pavement laying pattern(s), color(s), and texture.

5. This area shall be used as the standard by which the work is judged.

6. Subject to acceptance by the owner, mock up may be retained as part of the finished work.

7. If mock up is not retained, then remove and properly dispose

1.05   Delivery, Storage and Handling

A. **General:** Comply with Division 1 Product Requirement Section

B. **Deliver concrete grid units to the site in steel banded, plastic banded, or plastic wrapped packaging capable of transfer by forklift or clamp lift.** Unload grids at job site in such a manner that no damage occurs to the product or existing construction.

C. **Cover sand with waterproof covering to prevent exposure to rainfall or removal by wind.** Secure the covering in place.

D. **Coordinate delivery and paving schedule to minimize interference with normal use of buildings adjacent to paving.**

1.06   Environmental Conditions

A. **Do not install bedding materials or grid units during heavy rain or snowfall.**

B. **Do not install bedding materials and grid units over frozen base materials.**

C. **Do not install frozen bedding materials.**

1.07   Grid Maintenance Materials

A. **Supply [___] m² [(sf)] of [each type and color of grid unit] in unopened pallets with contents labeled. Store where directed.**

B. **Materials are from the same production run as installed materials.**

PART 2—Products

2.01   Concrete Grid Units

A. **Manufacturer:** Specify ICPI member manufacturer name

1. Contact: Specify ICPI member manufacturer contact information

B. **Concrete grid units:** Including the following:

1. Grid unit type: [Specify name of product group, castellated, lattice, etc.]
    - Material standard: Comply with material standards set forth in ASTM C1319

- Color and finish: [Specify color] [Specify finish]
- Color Pigment Material Standard: Comply with ASTM C979
- Size: [Specify] mm [([Specify] in.)] x [Specify] mm [([Specify] in.)] x [Specify] mm [([Specify] in)] thick

C. Manufactured in a plant where paving products are certified by ICPI as having met ASTM requirements in this specification.

2.03 Product Substitutions

A. **Substitutions:** No substitutions permitted

2.04 Bedding Materials

Note: If openings are filled with topsoil, use sand bedding. If the openings are filled with open-graded aggregate for additional runoff storage, the same or similar sized aggregate should be used for the bedding.

A. **General:** Sieve analysis per ASTM C136

B. **Bedding sand**

Note: The type of sand used for bedding is often called concrete sand. Sands vary regionally. Contact contractors local to the project and confirm sand(s) successfully used in previous similar applications.

1. Washed, clean, hard, durable crushed gravel or stone, free from shale, clay, friable materials, organic matter, frozen lumps and other deleterious substances
2. Conforming to the grading requirements in **Table 5-1**
3. Do not use limestone screenings or stone dust.

–OR–

B. **Washed, open-graded stone**

1. Conforming to the grading requirements in **Table 5-2**

Note: Finer gradations such as ASTM No. 89 stone may be used.

2.04 Fill Materials for Grid Openings

A. **Topsoil: Conform to ASTM D5268**

Note: Consult with local turf grass specialists for recommendations on grass seed mixture or sod materials.

B. **Grass seed (sod):** Mixture and source

–OR–

A. **Open-graded aggregate**

B. **Conforming to the gradation requirements in Table 5-2.** Do not use gravel.

2.04   Edge Restraints

A.   **Provide edge restraints installed around the perimeter of all concrete grid paving unit areas as follows:**

    1.   Manufacturer: [Specify manufacturer]

    2.   Material: Plastic, concrete, aluminum, steel, precast concrete, and cut stone

    3.   Material standard: [Specify material standard]

2.05   Accessories

A.   **Provide accessory materials as follows:**

    1.   Geotextile

        ■ Material type and description: [Specify material type and description]

        ■ Material standard: [Specify material standard]

        ■ Manufacturer: [Acceptable to concrete grid unit manufacturer] [Specify manufacturer]

PART 3—Execution

3.01   Acceptable Installers

A.   **[Specify acceptable paving subcontractors]**

3.02   Examination

Note: Compaction of the soil subgrade is recommended to a minimum of 95% standard Proctor density per ASTM D698 or AASHTO T-99 for pedestrian and vehicular areas. Stabilization of the subgrade and/or base material may be necessary with weak or saturated soil subgrades.

Note: Local aggregate base materials typical to those used for highway flexible pavements are recommended or those conforming to ASTM D2940. Compaction of aggregate is recommended to no less than 95% Proctor density in accordance with ASTM D698 for pedestrian and vehicular areas. Mechanical tampers are recommended for compaction of soil subgrade and aggregate base in areas not accessible to large compaction equipment. Such areas can include that around lamp standards, utility structures, building edges, curbs, tree wells, and other protrusions. The recommended base surface tolerance after compaction should be ±10 mm (±0.4 in.) over a 3 m (10 ft) straight edge.

Note: The elevations and surface tolerance of the aggregate base determine the final surface elevations of concrete grids. The installation contractor cannot correct deficiencies in the base surface with additional bedding materials. Therefore, the surface elevations of the base should be checked and accepted by the architect, engineer, general contractor or other designated party with written specification compliance certification to the paving subcontractor prior to placing bedding materials and concrete grids.

A.   **Acceptance of site verification conditions**

    1.   Contractor shall inspect, accept, and verify in writing to the grid installation subcontractor that site conditions meet specifications for the following items prior to installation of bedding materials and concrete grid units:

a. Verify that drainage and subgrade preparation, compacted density and elevations conform to specified requirements.

b. Verify that geotextiles, if applicable, have been placed according to drawing and specifications.

c. Verify that base materials, thickness, [compacted density,] surface tolerances, and elevations conform to specified requirements.

d. Provide written density test results for the soil subgrade, base materials to the owner, contractor, and grid installation subcontractor.

2. Do not proceed with installation of bedding materials and concrete grids until soil subgrade and base conditions are corrected by the contractor or designated subcontractor.

3.03 Preparation

A. **Verify that base is dry and certified by the contractor as meeting material, installation, and grade specifications and that geotextiles are ready to support sand, edge restraints, grids, and imposed loads.**

B. **Edge restraint preparation**

1. Install edge restraints per the drawings and manufacturer's recommendations at the indicated elevations.

2. Secure directly to compacted, finished base. Do not install on bedding sand.

3. The minimum distance from the outside edge of the base to the spikes shall be equal to the thickness of the base.

3.04 Installation

A. **Spread the sand No. 8 stone evenly over the compacted, dense-graded base course and screed uniformly to 13 to 25 mm (0.5 to 1 in.).** Place sufficient sand [stone] to stay ahead of the laid grids.

B. **Protect the grid units from foreign materials before installation.**

C. **Lay the grid units on the bedding sand in the pattern(s) shown on the drawings.** Maintain straight joint lines.

D. **Joints between the grids shall not exceed 5 mm (0.2 in.).**

E. **Fill gaps at the edges of the paved area with cut grids or edge units.**

F. **Cut grids to be placed along the edge with a double-bladed splitter or masonry saw.**

G. **Sweep topsoil No. 8 aggregate into the joints and openings until full.**

H. **Sweep the grid surface clear prior to compacting.**

I. **Compact and seat the grids into the screeded bedding sand No. 8 aggregate using a low-amplitude, 75-90 Hz plate compactor capable of at least 22 kN (5,000 lbf) centrifugal compaction force.** Use rollers or a rubber or neoprene pad between the compactor and grids to prevent cracking or chipping. Do not compact within 2 m (6 ft) of the unrestrained edges of the grid units.

J. **All work to within 2 m (6 ft) of the laying face must be left fully compacted at the completion of each day.**

K. **Broadcast grass seed at the rate recommended by seed source. Place sod plugs into openings.** Add topsoil to the surface to cover the seeds.

L. **Remove excess topsoil ASTM No. 8 aggregate on surface when the job is complete.**

M. **Distribute straw covering to protect germinating grass seed (sod).** Water entire area. Do not traffic pavement for 30 days if seeded.

3.05   Field Quality Control

A. **After removal of excess topsoil/aggregate, check final elevations for conformance to the drawings.** Allow 3 to 6 mm (0.12 to 0.24 in.) above specified surface elevations to compensate for minor settlement.

B. **The final surface tolerance from grade elevations shall not deviate more than ±10 mm (±0.4 in.) over a 3 m (10 ft) straightedge.**

C. **The surface elevation of grid units shall be 3 to 6 mm (0.12 to 0.24 in.) above adjacent drainage inlets, concrete collars or channels.**

D. **Lippage:** No greater than 3 mm (0.12 in.) difference in height between adjacent grid units.

3.06   Protection

A. **After work in the section is complete, the general contractor shall be responsible for protecting work from damage due to subsequent construction activity on the site.**

Recommended Outline for Specifications: Porous Unit Paving—Plastic Grids with Grass

PART 1—GENERAL

1.01   Summary

A. **Section**

   1.   Compacted aggregate road base

   2.   Plastic grid products

   3.   Grass sod or hydroseeding

B. **Related sections**

   1.   Section 31 20 00 Earth Moving

   2.   Section 33 40 00 Storm Drainage Utilities

   3.   Section 32 80 00 Irrigation

1.02   References

A. **American Society of Testing Materials (ASTM)**

   1.   C136 *Method for Sieve Analysis for Fine and Coarse Aggregate*

   2.   D698 *Standard Test Method for Laboratory Compaction Characteristics of Soil Using Standard Effort* 600 kN-m/m^3 (2,000 ft-lbf/ft^3)

   3.   D2940 *Standard Specification for Graded Aggregate Material for Bases or Subbases for Highways or Airports*

   4.   D 5268 *Specification for Topsoil Used for Landscaping Purposes*

1.03   Submittals

A. **Manufacturer's product data and installation instructions**

B. **300 x 300 mm (12 x 12 in.) section of plastic grids**

C. **Sieve analysis per ASTM C136 for the base course and sand fill materials**

1.04   Quality Assurance

A. **Installation by skilled workpeople with satisfactory record of performance on landscaping or paving projects of comparable size and quality.** Contractor shall provide documentation prior to starting work.

B. **Single-source responsibility:** Obtain each type of grid and fill materials from single sources with resources to provide products and materials of consistent quality, appearance, and physical properties without delaying progress of the work.

C. **Regulatory requirements and approvals:** Specify applicable licensing, bonding, or other requirements of regulatory agencies.

D. **Mock-up**

1. Locate where directed by the architect or engineer

2. Notify architect or engineer in advance of dates when mock-ups will be erected

3. Install minimum 10 m² (108 sf) of plastic grid units and topsoil or sod

4. Use this area to determine the quality of workmanship to be produced in the final unit of Work

5. This area shall be used as the standard by which the work is judged

6. Subject to acceptance by the owner, mock up may be retained as part of the finished work

7. If mock up is not retained, remove and properly dispose of

1.05   Delivery, Storage and Handling

A. **Protect plastic grid units from damage during delivery.**

B. **Store under tarp to protect from sunlight if time from delivery to installation exceeds one week.**

C. **Store grass seed in a dark and dry location.**

D. **When stored on the job site, cover topsoil piles with secure waterproof material to prevent erosion.**

E. **When stored on the job site, maintain sod in rolls and keep moist. Do not allow sod to dry out.**

1.06   Environmental Conditions

A. **Review installation procedures and coordinate plastic grid installation with other affected work.**

B. **Complete the paving of all adjacent hard surfaces prior to installing plastic grids and grass.**

Note: Gradients for plastic grid surfaces with grass vary from flat to 12% depending upon vehicle types using the surface. Fire lanes or surfaces used by other emergency vehicles will generally require a gradient of less than 6%.

C. **Do not use frozen materials or materials mixed or coated with ice or frost.**

D. **Exercise caution in handling plastic grid rolls in temperatures below 10°C (50°F).**

Note: Connectors in plastic grids can become stiff and can separate in cold weather. Individual grid units will retain the roll curl until warmed to room temperature. This can be aided by placement in sun for 15 to 20 minutes. If cold weather installation is anticipated, use plastic grids supplied in flat sheets that measure 1 x 1 m (40 x 40 in.).

E. **Do not start installation on frozen or saturated soil subgrade or aggregate base.**

1.07   Grid Maintenance Materials

A. **Supply** [___] m² [(sf)] of [plastic grid unit] in unopened pallets with contents labeled. Store where directed.

PART 2—Products

2.01   Plastic Grids

A. **Manufacturer:** Specify product name, manufacturer name and contact information

B. **Local sales representative:** [Specify name and contact information]

C. **Plastic grid paving units for grass:**

   1. [List material description, weight, etc.]

2.02   Aggregate Base Course

A. **Gradation conforming to state or local specifications for aggregate bases used under asphalt road surfacing.**

Note: If there are no state or local gradation specifications available, the material shall conform to the grading in ASTM D2940.

B. **Base course shall have a pH of 6.5 to 7.2 for turf root zone development.**

Note: Alternative base course materials such as crushed shell, limerock, and/or crushed lava may be considered for base course use, provided they are mixed with sharp sand (33%) and compacted.

2.03   Turf Materials

Note: Choose one of the following to suit project requirements.

A. **Washed concrete sand conforming to AASHTO M6 or ASTM C33**

–OR–

B. **United States Golf Association (USGA) "Green Section Recommendations For A Method of Putting Green Construction, Step 4—The Root Zone Mixture"**

C. **Topsoil:** Conforming to ASTM D5268

D. **Grass**

Note: Wear-resistant grass species generally include bluegrass/rye/fescue mix used for athletic fields in northern climates and zoysia, fescue, or Bermuda types in southern climates. Check with local sod and seed suppliers for preferred mixtures. Dedicated fire lanes can use the same grass species used on surrounding turf. Parking applications require high wear-resistant species, generally planted by seed or sprigging. Select sod or seed below to suit project requirements.

   1. **Sod**: 13 mm (0.5 in.) thick (soil thickness) rolled from a reputable local grower. Wear-resistant grass species, free from disease and in excellent condition, grown in sand or sandy loam soils only. Do not use sod grown in clay, silt, or high organic soils such as peat.
      –OR–

2. **Seed**: Use certified species recommended by seed suppliers, considering local environmental conditions and expected traffic. Provide in containers with labels indicating seed name(s), lot number, net weight, percent weed seed content, and guaranteed percent purity and germination.

   - Mulch: Wood or paper cellulose types used in hydroseeding operations. Do not use of straw, pine needles, mulches, etc.

   - Fertilizer: A commercial starter type with guaranteed analysis of 17-23-6 (nitrogen-phosphorus-potassium) or as recommended by local grass seed supplier for rapid germination and root development.

Note: Fire lane lanes must be identified regarding their entrance and physical location with the placement of signs, gates, curbs, bollards, etc. Specific signage wording and other details should be coordinated with and approved by the local fire department.

Part 3—Execution

3.01   Inspection

Note: Verify with the local fire department if certificates of inspection are required. For fire lane projects, fire department inspectors should inspect the installation during soil subgrade compaction, installation of the base course, and the plastic grids. Most small projects can accommodate these inspections all on the same day.

A. **Confirm soil subgrade compaction test results comply with specifications.** Verify that the soil subgrade is free of debris and the gradients and elevations conform to the drawings. Do not start base and plastic grid installation until unsatisfactory conditions are corrected.

B. **Installation of base and plastic grids constitutes acceptance of prepared soil subgrade and responsibility for satisfactory performance.** If existing conditions are found unsatisfactory, contact the project manager for resolution.

3.02   Preparation

Note: The compacted soil subgrade should be structurally adequate to receive designed base course, wearing course, and designed loads. Generally, excavation into undisturbed soils requires no additional modification. Fill soils and otherwise structurally weak soils may require modifications such as geotextiles, geogrids, and/or stabilization. Confirm that grading and soil porosity provides adequate subsurface drainage.

A. **Place aggregate base course material over compacted soil subgrade to grades shown on the drawings.** Place maximum 150 mm (6 in.) lifts, compacting each lift separately to a minimum of 95% standard Proctor density per ASTM D698 or AASHTO T99. Leave minimum 25 mm (1 in.) to 40 mm (1.57 in.) for the plastic grids and sand/sod fill to the finished grade. Surface tolerance of the compacted aggregate base shall be ±10 mm (0.4 in.) over a 3 m (10 ft) straightedge.

3.03   Plastic Grid Unit Installation

A. **Install the plastic grids by placing units with rings facing up, using pegs and holes provided to maintain proper spacing and interlock the units.** Shape units with pruning shears or knife as needed.

B. **Units placed on curves and slopes shall be anchored to the base course, using nails with fender washers.** Follow plastic grid manufacturer's instructions on fender washer size, nail size and spacing.

Note: Edit items C through G below if sand is not used in the grid openings and revise for use of topsoil only.

C. **Install sand in rings as they are placed in sections directly from a dump truck or from buckets mounted on tractors.**

D. **Dump trucks and tractors shall exit the site by driving over rings filled with sand.**

E. **Spread the sand from the pile using flat bottomed shovels and/or wide asphalt rakes to fill the rings.**

F. **Use a broom for final finishing of the sand.** Settle the sand into the plastic grids with water from a hose, irrigation heads, or rainfall.

G. **The finish grade shall be at the top of rings and no more than 6 mm (0.25 in.) above top of the plastic grids.**

3.04   Installation of Grass

Note: Select items below to meet grass installation method desired. Hydroseeding is the preferred grass installation method. If manual seeding is used, revise item A below to include the application rate for the grass seed.

A. **Use hydroseeding/hydro-mulching with a combination of water, seed, and fertilizer homogeneously mixed in a purpose-built, truck-mounted tank.** Spray the seed mixture onto the plastic grids at rates recommended by the hydroseeding manufacturer with uniform coverage.

B. **Following germination of the seed, immediately re-seed areas lacking germination larger than 20 x 20 cm (8 x 8 in.).** Fertilize and keep turf plants moist during development.

–OR–

A. **Install sod directly over sand-filled rings.** Place sod strips with each compressed against the next strip to form tight joints.

B. **Fertilize and keep sodded areas moist during root establishment for a minimum of 3 weeks.**

C. **Protect sodded areas from any traffic other than emergency vehicles for 3 to 4 weeks or until the root system has penetrates below the plastic grid units.**

-OR-

A. **Install grass seed at rates recommended on the grass seed packaging or from the supplier.** Apply a commercial topsoil mix conforming to ASTM D5268 approximately 25 mm (1 in.) thick to aid germination. Fertilize and keep seeded area moist during grass development.

3.05   Protection

A. **Protect partially completed paving against damage from other construction traffic when work is in progress.** Any barricades constructed shall be accessible by emergency and fire equipment during and after installation.

B.   **Protect adjacent work from damage during base and plastic grid installation.**

Note: Select one item below to conform to the grass installation method.

C.   **Protect seeded areas from any traffic, other than emergency vehicles, for 4 to 8 weeks or until the grass is mature and can adequately resist damage from vehicular traffic.**

–OR–

C.   **Protect sodded areas from any traffic, other than emergency vehicles, for a period of 3 to 4 weeks or until the root system has penetrated below the plastic grid units.**

3.06   Cleaning

A.   **Remove and replace segments of plastic grid units where three or more adjacent rings are broken or damaged, reinstalling each per this section so no evidence of replacement is visible.**

B.   **Remove all excess materials, debris and equipment from site when work is complete.** Repair any damage to adjacent materials and surfaces resulting from installation of this work.

5.9   General Resources

Interlocking Concrete Pavement Institute (ICPI) (1999). *ICPI Tech Spec 8 Concrete Grid Pavements*, Interlocking Concrete Pavement Institute, Herndon, VA, <www.icpi.org>.

Ferguson, B. (2005). *Porous Pavements*. CRC Press, Boca Raton, FL.

Websites:

Interlocking Concrete Pavement Institute (ICPI) PICP resources for design, construction and maintenance <www.icpi.org>.

Invisible Structures <www.invisiblestructures.com>.

Low-Impact Development Center <www.lowimpactdevelopment.org>.

North Carolina State University <www.bae.ncsu.edu/info/permeable-pavement>.

5.10 References

Booth, D. B. and Leavitt, J. (1999). "Field Evaluation of Permeable Pavement Systems for Improved Stormwater Management." *Journal of the American Planning Association*, 314–325.

Brattebo, B. O., and Booth D. B. (2003). "Long-term Stormwater Quantity and Quality Performance of Permeable Pavement Systems." *Water Resources*, Elsevier Press, 37, 4369–4376.

Collins, K. A., Hunt, W. F., Hathaway, J. M. (2007). "Hydrologic and Water Quality Comparison of Four Types of Permeable Pavement and Standard Asphalt In Eastern North Carolina." Biological and Agricultural Engineering Department, North Carolina State University, Raleigh, NC.

Day, G. E. (1980). "Investigation of Concrete Grid Pavements," *Stormwater Management Alternatives*, J. T. Tourbier and R. Westmacott, eds.

Day, G. E., Smith, D. R. and Bowers, J. (1981). "Runoff and Pollutant Abatement Characteristics of Concrete Grid Pavements." Bulletin 135, Virginia Water Resources Research Center, Virginia Tech, Blacksburg, VA.

Goforth, G. F., Diniz, E. V. and Rauhut, J. B. (1984). "Stormwater Hydrological Characteristics of Porous and Conventional Paving Systems." *EPA report number 600283106*, Municipal Environmental Research Laboratory, Office of Research and Development, United States Environmental Protection Agency (US EPA), Cincinnati, OH.

Smith, D. R. (1981). "An Experimental Installation of Grass Pavement—IV." *Life Cycle Cost and Energy Comparison of Grass Pavement with Asphalt Based on Data and Experience with the Green Parking Lot*, City of Dayton, OH.

Smith, D. R. and Hughes. M. K. (1981). "An Experimental Installation of Grass Pavement—I." *Project Description*, City of Dayton, OH.

Smith, D. R. (1984). "Evaluations of Concrete Grid Pavements in the United States." *Proc., 2nd International Conference on Concrete Block Paving*, Technical University of Delft, Delft, The Netherlands, 330–336.

University of Delaware Water Resources Center (1980). University of Delaware Water Resources Center, Newark, DE.

6 Alternative Technologies

6.1 Description of Products

In recent years, new permeable pavement product variations have entered the market. This chapter describes general characteristics for several emerging products. The following types of permeable pavements are discussed:

- Pervious pavers
- Rubber overlay pavement
- Rubber composite permeable pavers
- Engineered aggregates or matrices

6.2 Pervious Pavers

Pervious paver pavement consists of unit paving typically made of a combination of uncrushed or crushed stones bound together with a polymer (**Figure 6-1**) or cement (**Figure 6-2**). These pavers differ from permeable interlocking concrete paver (PICP) systems in that the pavers themselves are permeable and, as such, the permeability is related to the entire paver surface and not limited to the open joints between them. Pervious pavers generally are rectangular or square and can be manufactured in custom shapes and sizes. The paving units can be made with different aggregate sizes and colors. Other variations include concrete mixes with coarse sand and pervious ceramic paving units. Depending on the intended use and specific design goals, pervious pavers may be situated over a permeable aggregate base/subbase or be installed directly over the underlying soil (without a base/subbase).

Figure 6-1
Example of a pervious paver bound with a polymer Xeripave®
Source: CH2M HILL

Figure 6-2
Example of a pervious paver bound with cement
Source: Bayshore Concrete

For polymer and cement bound pavers, the design typically consists of pea-sized stone that is formed into a solid matrix with a binder creating void spaces to allow water to pass. The paver matrix typically contains 20% to 40% voids and is characterized by high surface infiltration rates when new. The paving units vary in thickness. A 50 mm (2 in.) minimum thickness for polymer bound units can be used for areas of light traffic load, including pedestrian areas and areas subject to bicycles, wheelchairs, and golf carts. With the proper base/subbase depth specifically designed to support the required loads, the pavers can support periodic use by emergency or utility vehicles. Paver thickness can be increased up to 100 mm (4 in.) for use in heavier or higher frequency traffic areas, including parking lot entrance and exit areas. In all cases, a site-specific design load analysis should be conducted by a qualified engineer or designed/guaranteed by a manufacturer to determine the design criteria for the base/subbase material.

Cement-bound pervious units also vary in thickness depending on the particular manufacturer and anticipated traffic load. Thicker pavers are generally recommended for heavier traffic areas. There are cement bound units that have reduced sand content and certain units are manufactured with almost no sand. These units are typically suitable for pedestrian and residential driveway applications. They do not meet ASTM requirements for units used in permeable interlocking concrete pavements. The design engineer should verify the exact material makeup of the paver when designing a permeable system and compare the requirements with expected load/use.

As with other types of permeable pavement, annual or semi-annual maintenance is typically recommended with a sweeper/mechanical vacuum. Specific manufacturers may recommend power washing to remove accumulated sediments lodged in the surface. The type of cleaning and frequency will be product and site-specific. In areas where fines, airborne dust, or debris are excessive, more frequent maintenance may be required.

For pervious natural aggregate pavers bound with polymers, a manufacturer recommended grout should be used between pavers in areas subject to vehicular loading. If grout is not used, abrasion may result between pavers and paver breakdown. In pedestrian applications, pavers may be installed edge to edge without grouting, if approved by the manufacturer and design engineer.

Pervious Paver Performance, Concerns and Limitations

Pervious pavers have flow-through rates as high as 1,300 cm/hr (5,000 in./hr) when new. This rate is typically higher than that of other permeable pavements. This high flow-through rate is a result of the large void space within the paver.

Cement-bound pervious pavers can be subject to a higher risk of damage from deicing materials. Deicing materials can break down the cement and cause surface spalling or disintegration of the matrix. This concern also applies to pervious concrete and is further discussed in Chapter 3. Solid concrete pavers in permeable applications (i.e., PICP) are generally at a much lower risk from this type of damage since they have a lower exposed cement surface area.

For polymer bound units, the binders chosen are typically environmentally inert. An advantage of these polymer bound units is that they are typically not affected by freeze/thaw, heat, deicing materials, or UV, and demonstrate resistance to petroleum (automotive) fluids.

Pervious Paver Design and Installation Considerations

Pervious pavers are typically installed over a 2.5 to 4 cm (1 to 1.5 in.) thick bedding, typically clean, washed AASHTO No. 8 stone overlying a 10 to15 cm (4 to 6 in.) thick layer of clean, washed, well-compacted aggregate, typically AASHTO No. 57 stone. In areas where native soils allow for very slow infiltration of runoff, a deeper layer of stone base (and greater volume of stone storage) may be required. Also, a subbase of clean, washed aggregates of larger gradation (i.e., AASHTO No. 2 stone) may be included in designs for vehicular areas. If desired by the designer, geotextile can be used for structural support between the underlying soils and clean crushed stone material or as a separator between aggregate layers. Sand is never recommended for bedding or jointing pervious pavers as this may cause differential settlement or result in a decline in surface infiltration rates. Pavers come precast, so no surface cure time is required during installation. **Figure 6-3** illustrates a typical cross section of a pervious paver system.

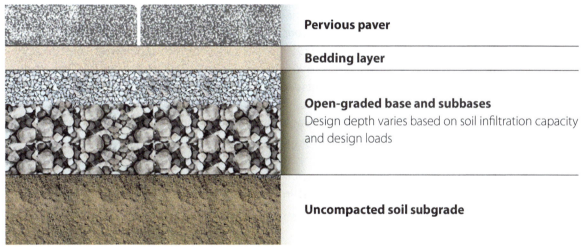

Pervious paver

Bedding layer

Open-graded base and subbases
Design depth varies based on soil infiltration capacity and design loads

Uncompacted soil subgrade

Figure 6-3
Generic cross section for a pervious aggregate paver system
Source: © VHB

Pervious Paver Operation and Maintenance

The maintenance requirements are similar to that of other pavement types and are described in **Chapter 8**.

Pervious Paver Cost Information

Typical material costs for pervious aggregate pavers (2013) are as follows:

- 30.5 cm x 30.5 cm x 5 cm (12 in x 12 in x 2 in) pavers—$7 to $8 each
- 40.6 cm x 40.6 cm x 5 cm (16 in x 16 in x 2 in) pavers—$13 to $15 each
- 30.5 cm x 30.5 cm x 10 cm (12 in x 12 in x 4 in) pavers—$13 to $15 each

These costs do not include installation costs, which vary by region and contractor experience. Installation costs are usually an additional $0.09 to $0.19 m² ($1 to $2/sf), not including the cost of base/subbase and subgrade materials and preparation.

Pervious Paver Specifications

Currently, there are no ASTM standards for pervious pavers. For detailed product and installation information, designers should refer to manufacturer's recommendations for testing specific products and for developing specifications. The design engineer should get direct input from the manufacturer based on the specific use and application.

6.3   Rubber Overlay Pavement

Rubber overlay pavements are a type of permeable pavement made from a mix of recycled rubber granules, dry aggregate, and a proprietary binder. The rubber granules are typically 0.6 to 1 cm (0.24 to 0.4 in.) in size. When bound together into the overlay matrix, they form a permeable surface with open voids. If properly mixed, the binder enables the rubber to resist degradation from transmission fluid, brake fluid, hydraulic fluid, salt water, oil, chlorine, ozone, bromine, muriatic acid, and other hostile materials. The pavement material is typically poured in place by hand using specialized labor and equipment. Rubber overlay pavement is appropriate in light traffic and low speed (<40 kph [<25 mph]) applications such as pathways, courtyards, driveways, horse barns, islands (traffic, landscape, parking), median strips, parking lots, sidewalks, playground areas, and tree surrounds. Rubber overlay pavements are available in a range of colors (**Figures 6-4 and 6-5**).

Rubber overlay pavements have flow through rates as high as 1,000 cm/hr (400 in./hr). These types of pavements demonstrate elastic properties that help the product resist cracking.

Some manufacturer's literature describes product applications installed over impermeable bases (existing asphalt, concrete pavers, and compacted aggregate). In these instances, this product would not function as a standard permeable pavement.

The following information pertains only to rubber overlay pavements installed over permeable base/ subbases. For information on overlays over impermeable surfaces, see **Chapter 2, Section 2.9 Resources for Specifications**.

Figure 6-4
Example of a rubber overlay pavement in a landscaping application
Source: PorousPave, Inc.

Figure 6-5
Example of rubber overlay pavement in a driveway application
Source: PorousPave, Inc.

Rubber Overlay Pavement Performance and Limitations

Manufacturer information indicates that rubber overlay pavement is intended for pedestrian and low speed (<40 kph [<25 mph]) traffic uses. It is not intended for heavy vehicular traffic applications. If placed over a dense-graded base, it will not be a permeable pavement. Rubber overlay pavements are slip resistant and ADA compliant.

Rubber Overlay Pavement Design and Installation Considerations

Rubber overlay pavements are poured in place directly over a level pavement subbase and require a 24-hour cure time. The thickness of the overlay pavement depends on the particular pavement use. Foot traffic installations generally require a 38 mm (1.5 in.) thick overlay and a minimum 50 mm (2 in.) aggregate base. Vehicle use applications generally require a 50 mm (2 in.) overlay and a minimum 100 to 150 mm (4 to 6 in.) aggregate base. Manufacturers should be consulted for more specific design and installation requirements.

Rubber Overlay Pavement Operation and Maintenance

The maintenance requirements are similar to that of other pavement types and are described in Chapter 8.

Rubber Overlay Pavement Cost Information

The cost of rubber overlay pavements varies and is dependent on total system requirements and location. Typical material costs for 50 mm (2 in.) of overlay pavement surface range from $0.93 to $1.21/m^2 ($10 to $13/sf) installed. This cost does not include the cost of base/subbase and subgrade materials and preparation.

Rubber Overlay Pavement Specifications

Currently, there are no ASTM standards for rubber overlay pavements. A certified installer is recommended for product installation. For detailed product and installation information, designers should refer to manufacturer's recommendations for specific products and test methods.

6.4   Rubber Composite Permeable Pavers

Rubber composite pavers are a light-weight paver made from up to 95% post-consumer recycled materials, such as scrap tires and plastic (Figure 6-6). The pavers can be placed with open joints between them and backfilled with small aggregate to create a permeable surface. The pavers themselves are not permeable. Approximately 9.3 m^2 (100 sf) of rubber composite pavers will typically contain 50 truck tires and 1,500 gallon-sized plastic containers. The pavers are roughly one-third the weight of standard concrete pavers and are available in a variety of shapes and sizes. The composite grids and pavers can be easily cut to specific sizes and shapes.

Figure 6-6
Rubber composite paver and grid
Source: AZEK® Pavers

Rubber composite pavers are suitable for pedestrian and light automobile traffic. Typical applications include driveways, patios, sidewalks, playground areas, and pathways.

Manufacturer's literature describes product applications installed without open surface joints and/or applications installed over impermeable bases (e.g., existing asphalt, concrete pavers, and compacted aggregate). In these instances, this product would not function as a standard permeable pavement. The following information pertains only to rubber composite permeable pavers installed with open joints and placed over permeable bases/subbases.

Rubber Composite Permeable Paver Performance, Concerns, and Limitations

Rubber composite pavers do not absorb moisture, making them stain-resistant and unaffected by freeze/ thaw damage such as cracking, spalling, and chipping. The pavers are also unaffected by salt and most other deicers and household chemicals. The recycled rubber and plastic materials used in these products do not break down naturally and have shown long-term durability. No sealing or staining is required.

Tests have indicated that tire rubber can leach small amounts of zinc into underlying soils. While rubber composite pavers contain recycled tire rubber, very little of the rubber is typically in contact with the underlying base/subbase since the composite material encapsulating the rubber is almost entirely plastic. It is assumed that the risk of leaching is low; however, additional research is needed to fully assess the risk (AZEK 2013).

Rubber composite pavers are not recommended for use near or around fire pits due to the potential for slow flame spread and surface scarring.

Rubber Composite Permeable Paver Design and Installation Considerations

Rubber composite permeable pavers are easy to install due to the use of a grid system that takes the difficulty out of manually setting and aligning concrete pavers. Once placed, the open surface joints should be backfilled using a clean, washed aggregate, typically AASHTO No. 8 stone. The pavers are one-third the weight of concrete and can be cut with traditional wood saws. Since the pavers are precast, no surface cure time is required during installation. Not including base/subbase preparation, typical installation time for rubber composite paver is reduced by 60% compared to standard concrete pavers because they are lightweight and easy to transport and cut.

Rubber composite permeable pavers are typically installed over an open-graded base/subbase. The stone sizes in open-graded bases/subbased can be as large as 76.2 mm (3 in.) and as small as 6 mm (0.24 in.). A solid edge restraint is typically used to hold the pavers in place. The typical cross section is similar to that shown in **Figure 6-3**. However, due to their non-ASTM standard status, the designer will require specific manufacturer input to prepare details for the specific use and site conditions.

Rubber Composite Permeable Paver Operation and Maintenance

Rubber composite permeable pavers require occasional replenishment of joint fill and surface cleaning using a vacuum to remove accumulated sediment. Chemical spills (e.g., petroleum, paint, etc.) should be cleaned as soon as possible with an appropriate industrial cleaner and rinsed thoroughly with water.

Chemical spills or stains left untreated over long periods of time may degrade the surface finish and integrity of the product and would likely not be covered under warranty.

Rubber Composite Permeable Paver Cost Information

Rubber composite permeable pavers cost an average of to $75 to $108 m² ($7 to $10/sf) for the pavers and installation grid, although actual costs vary by region and the size of the installation. Since the product is lightweight, easy to transport, easy to cut and quick to install, labor costs for installation are generally low. Fully installed, rubber composite pavers generally range from $108 to $129 m² ($10 to $12/sf), including surface materials and labor. This cost does not include the cost of base/subbase and subgrade materials and preparation.

Rubber Composite Permeable Paver Specifications

Currently, there are no ASTM standards for rubber composite permeable pavers. For detailed product and installation information, designers should refer to manufacturer's recommendations for specific products and test methods.

6.5   Engineered Aggregates or Matrices

Engineered aggregates or matrices refer to materials that can be applied to permeable pavement systems to elevate water quality treatment performance. These materials are designed to provide redundant pollutant removal by adsorbing dissolved-phase pollutants (i.e., phosphorus, metals, etc.). When using these materials as a component of a permeable pavement system, the intent is to maintain the same permeability and filtering capability of standard permeable pavements, with the added benefit of target pollutant treatment to protect surface and groundwater resources. Some materials can be applied as admixtures during the manufacture process of pervious materials, creating a treated matrix of the permeable pavement materials, which increases the adsorptive functionality of the materials themselves (Sansalone 2004 and 2008).

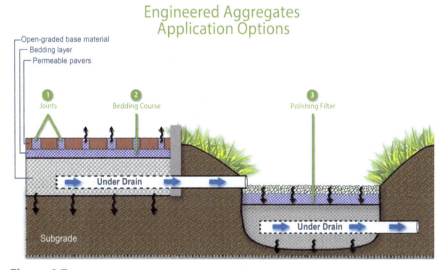

Figure 6-7

**Typical applications for Engineered Aggregates (1) in pavement joints,
(2) in pavement bedding course or (3) as an added polishing system.**
Source: Imbrium Systems Corp.

Engineered aggregates can be applied in a variety of applications, such as permeable pavement bedding, joint fill or can be separately applied downstream of the permeable pavement as a polishing water quality treatment cell (see **Figure 6-7**). Certain products may be applicable as an admixture within the manufacturing process of the permeable pavement itself. **Figure 6-8** shows a pervious concrete application with an added sorbative add mixture.

Figure 6-8
Example of pervious concrete installation with an added sorbtive admixture
Source: Imbrium Systems Corp..

Engineered aggregates or matrices may be most appropriately applied as an additional component to permeable pavement systems to elevate total pollutant removal, further improve water quality for impaired watersheds, address total maximum daily loads requirements (TMDLs) or improve consistency of the infiltrated effluent.

Application of this treatment technology may be of the highest interest within watersheds, which already have surface water impairments, existing TMDLs or forthcoming TMDLs requiring advanced treatments, and pollutant load reductions beyond standard filtration non-engineered aggregates and soils. This may have limited adsorptive properties for critical pollutants.

The desired water quality enhancement from using engineered aggregates or matrices would likely dictate the application, longevity, and cost.

Engineered Aggregates or Matrices Performance and Limitations

Water quality research conducted (see **Appendix B**) has generally indicated that water quality output data for certain pollutants is not yet consistent and can not be accurately predicted. However, utilizing well-selected engineering materials to enhance water quality benefits, to elevate the total pollutant removal, or improve consistency of the infiltrated effluent of permeable pavement system has significant potential.

Pollutants, such as nutrients, are known to vary in fate and transport. Their removal is a dynamic process, requiring more than physical filtration to achieve elevated performance and more consistent effluent quality. This was identified in field research (Madge 2004; Vaze 2004) suggesting that consistently higher levels of total phosphorus removal (>65%) are achieved by treatment practices removing particles as small as 11-microns and capturing/retaining a significant fraction of the dissolved phosphorus. In order to treat the dissolved phase of phosphorus, removal processes such as adsorption onto engineered aggregates, would need to be implemented. The use of adsorptive materials that may have potential to leach alternative pollutants such as nutrients (some natural organic materials), heavy metals (slag), or contribute to high pH (>9.0) should be avoided.

Two other parameters that must be considered in addition to breakthrough (i.e., exhaustion of pollutant capture that results in release) to assist performance are the possibilities of desorption (i.e., the release of the pollutant after it is captured) and leaching of alternative pollutants that may be inherent to the material being used. Some natural materials for example are known to be comprised of phosphorus. Over time, they may easily release or desorb this pollutant. Recycled media, often waste by-products and sometimes marketed as "green technology," may contain debris or potentially leach toxic pollutants that become apparent when tested in stormwater of typically low background pollutant concentration (Minton 2005). Many waste by-products may have the ability to adsorb a specific pollutant; however, due to the variable and often unknown material composition, they could leach alternative pollutants such as heavy metals or contribute a high alkalinity, raising the pH above acceptable discharge levels. These potential phenomena should be well understood and evaluated prior to selecting a material. Like soil, all engineered aggregates have a finite capacity to adsorb dissolved pollutants. The best determination of the longevity to perform in a given application is to evaluate breakthrough material characteristics and test results. This is described under **Section 6.5 Engineered Aggregates or Matrices**.

Engineered Aggregates or Matrices Design and Installation Considerations

Engineered aggregates are applied in a similar fashion to a permeable pavement system as standard aggregates. In general terms, a design that increases overall runoff contact time with engineered aggregates will likely lend to higher dissolved pollutant removal.

Material Selection

As a component integral to the permeable pavement system, materials used to adsorb dissolved pollutants should be specified based on physical characteristics as well as clear performance characteristics. There are three key items to consider when evaluating engineered aggregate material.

- **Material integrity and the gradation size of engineered materials**. Engineered aggregate physical properties, such as gradation size, should be characterized. The same basic requirements apply to provide strength, durability, and gradation size achieve or exceed the typical standards depending on the application.

- **Performance of engineered aggregate or matrices materials**. The performance of the material should be identified and understood prior to being implemented. At a minimum, calculating the amount of dissolved pollutant removed over a period of time through adsorption should be based on breakthrough performance testing under identifiable test conditions. The following parameters should be included as a minimum: solution influent/effluent concentrations, solution pH, solution surface loading rate, volumes, media parameters, and media depth. More rigorous performance analysis of engineered aggregates or matrices may be pursued, such as: in-depth testing of adsorption isotherm, reaction kinetics, breakthrough, and desorption (Wu 2008).

  Absorptive materials considered should be reviewed for potential unexpected leaching, alternative pollutants (i.e., heavy metals, nutrients or pH), and the potential for desorption of the pollutant of concern.

- **Breakthrough performance testing**. Breakthrough data is generated by continuous flow through testing. A test solution is continuously pumped through a media bed and samples are taken at regular intervals. Testing typically continues until the removal rate drops to a target percent removal or effluent level. Engineered aggregate or matrices lifetime predicted by this type of data is useful in calculating and determining long-term performance.

  Breakthrough performance testing is the most useful test parameter for properly sizing a system. Breakthrough data combines a practical value of an engineered aggregates or matrices sorption capacity with kinetic performance values that change, illustrating the diminishing capacity a media will experience over time. Breakthrough data provides a more realistic quantification of performance lifetime than equilibrium or kinetic data, which can easily be used to size and design systems using these materials. As water containing a dissolved pollutant flows through a bed of adsorptive media, the pollutant concentration changes as a function of depth in the media bed and time. The effluent concentration increases with time as the media adsorptive capacity is consumed, until "breakthrough" occurs and the effluent concentration has reached some predetermined percentage of the influent concentration or effluent level.

At breakthrough, the media will no longer provide the required dissolved pollutant removal, even though the contact time has not changed and the media is not completely saturated. Most theories used to describe this breakthrough behavior are based on an assumption that the adsorption process can be described as a first- or second-order reaction rate expression, and that the rate of this reaction is controlled by the remaining unused capacity of the media at a particular time. Using a model such as that proposed by Bohart (1920) or Thomas (1944), one can predict the performance of a media bed based on experiments conducted on a small-scale apparatus, assuming adequate information is known.

- **Other performance variables**. Performance variables should be taken into account to avoid large discrepancies between small-scale experimental results and that of large installations. These variables include media gradation size (which strongly affects the surface area available for adsorption reactions to take place upon), influent concentration, and surface loading rate (the flow rate of influent water per cross-sectional area of the filter media perpendicular to the direction of flow). Ultimately, all of these variables should be comparable to the engineered permeable pavement system and application for which they are intended so engineers can be provided with useful performance data to ensure a properly sized system with long-term performance.

Installation

Engineered aggregates can generally be installed in a similar fashion as standard aggregate, unless otherwise indicated by a supplier. Permeable pavement materials that have been manufactured with specialized admixtures should require no additional installation requirements, unless otherwise indicated by a supplier. **Figure 6-8** shows the installation of a pervious concrete surface with an added sorbtive admixture.

Engineered Aggregates or Matrices Operation and Maintenance

Engineered aggregates and matrices function to adsorb dissolved pollutants. These materials, like all adsorptive materials including natural soils, have a finite capacity to adsorb dissolved pollutants. As such, these materials shall be applied in a manner in which their adsorptive capacity is designed to outlast the useful life of the treatment requirements or as a structural pavement application. Alternatively these materials can be utilized in permeable pavement site design to allow for ease of removal and replacement of chosen materials.

Engineered Aggregates or Matrices Cost Information

Engineered aggregate costs are higher than standard aggregates since these materials often have been engineered and manufactured to provide an added pollutant removal value. The cost of admixtures is an added cost to a typical permeable pavement application. Generally a system cost increase of 5% to 20% should be anticipated. Factors that impact costs of engineered aggregates include volume of engineered aggregates used, location of project versus supply, application/permeable paver system design, and performance longevity (cost performance basis). If the cost of engineered aggregates or admixtures is compared from a performance basis (i.e., cost per pound of dissolved pollutant removed), the cost will typically be less expensive than standard pavements and standard aggregates.

Engineered Aggregates or Matrices Specification

For detailed product and installation information, designers should refer to manufacturer's recommendations for specific products.

The items below are some general items that could be used in specifications for engineered aggregates as a bedding layer underneath a permeable pavement system to provide elevated water quality performance through the removal of dissolved pollutants. These also address some concerns with the use of aggregates:

- A minimum layer depth of "x" centimeters of engineered aggregates, which do not desorb the pollutant of concern, nor shift the effluent pH greater than 9.0 shall be used to enhance the stormwater treatment practice and remove the dissolved pollutant phase. The engineered aggregates shall not leach alternative pollutants such as nutrients or heavy metals.

- The engineered aggregate layer shall be placed on top of the underlying aggregates, and the layer of engineered aggregate should be completely leveled during installation to allow even water flows through the engineered aggregate bed.

- Other bedding aggregates applied per local/state requirements to physically filter sediment and sediment-bound pollutants should not displace the engineered filtration media.

Below is an another example for specification language that may be considered for engineered aggregates used in a downstream surface filter to provide elevated water quality performance by removing dissolved pollutants from effluent infiltrated from a permeable pavement system:

- A volume of "X" cubic meters of engineered aggregates, placed within a filtration system at "y" centimeters of depth will be applied to remove dissolved pollutants. The engineered aggregates will not desorb the pollutant of concern, nor shift the effluent pH greater than 9.0 and shall be used to enhance the stormwater treatment practice and remove the dissolved pollutant phase. The engineered aggregates shall not leach alternative pollutants, such as nutrients or heavy metals.

- The engineered aggregate layer shall be placed on top the underlying layer, and the layer of engineered aggregate should be completely leveled during installation to ensure that water flows evenly through engineered aggregate bed.

- Other bedding aggregates applied per local/state requirements to physically filter sediment and sediment-bound pollutants should not displace the engineered filtration media.

6.6 References

Bohart, G. S., and Adams, E. Q. (1920). "Some Aspects of the Behavior of Charcoal with Respect to Chlorine." *Journal of American Chemical Society*, 42, 523–544 .

Minton, G. (2005). "Stormwater Treatment: Biological, Chemical Engineering Principles."

Sansalone J., Kuang, X., and Ranieri, V. (2008). "Permeable Pavement as a Hydraulic and Filtration Interface for Urban Drainage." *Journal of Irrigation and Drainage Engineering.* 134(5), 666.

Sansalone, J., and Teng, Z. (2004). "In-situ Partial Exfiltration of Rainfall Runoff. I: Quality and Quantity Attenuation." *Journal of Environmental Engineering.* 130(9), 990.

Thomas, H. C. (1944). "Heterogeneous Ion Exchange in a Flowing System." *Journal of American Chemical Society,* 66, 1664–1666.

Vaze, J., and Chiew, F. (2004). "Nutrient Loads Associated with Different Sediment Sizes in Urban Stormwater and Surface Pollutants." *Journal of Environmental Engineering*, American Society of Civil Engineers, 130(4), 391.

Wu, T., Gnecco, I., Berretta, C., Ma, J., and Sansalone, J. (2008). "Stormwater Phosphorus Adsorption on Oxide Coated Media." Water Environment Federation—WEFTEC.

7 Achieving Success and Avoiding Failures with Permeable Pavements

7.1 Introduction

This chapter provides experienced-based recommendations on how to achieve success with permeable pavements. This information is based on an informational survey completed by designers, contractors, owners, developers, and manufacturers working with various permeable pavements systems (EWRI 2010).

The four key areas presented are planning and site selection, design, construction, and operations and maintenance. Many of the concerns identified in this chapter have been mentioned in earlier chapters with the specifics for the individual types of pavement systems. This chapter provides insights and best practices to assist in the successful use of permeable pavement for stormwater management.

7.2 Planning and Site Selection

During the planning and site selection phase, key elements to consider include:

- **An integrated design team formed at the planning level**. A successful project will typically include the formation of an interdisciplinary design team (**Figure 7-1**) during the initial stages of site planning that continues to collaborate throughout all phases of the project. It is critical to engage civil and geotechnical engineers and hydrogeologists early on in the planning phases so they can provide existing conditions information and advice on site layout based on the physical conditions/constraints at the site. Promoting permeable pavement as a stormwater strategy can be challenging to implement if the engineering team is brought onto a project after the site layout has been completed. Geotechnical input regarding existing subsurface soil conditions (including permeability, contamination and subsurface geology) will influence permeable pavement design location options. This will further influence the site access and building and parking layouts for which the site/civil engineer will have input. A preliminary site layout prepared by these informed individuals can be evaluated by planners for conformance to local ordinances and local development policies as the layout progresses. Such findings can also guide the project architect and landscape architect regarding building location and site design. The developer can provide additional

Figure 7-1

A well-coordinated interdisciplinary design team is essential to a successful permeable pavement design
Source: SvR Design Company

guidance on construction sequencing and equipment use favorable to minimizing erosion and soil compaction. All of this collaborative activity should be framed in the context of reducing costs for the project owner.

- **Location, location, location**. Aside from site topography, hydrology, and soil conditions, other important factors to consider when selecting a location for permeable pavements are adjacent land uses and nearby landscape elements. These areas could be sources of stormwater sheet flow. Therefore, avoid locating permeable pavement adjacent to loose or erodible soils that might clog the permeable pavement surface with sediment. Facilities should also not be located in areas that could wash onto the pavement and contaminate the subsurface soils and groundwater, containing hazardous materials such as an industrial area, fuel station, or a car wash. In arid regions, the threat of windblown dust depositing on permeable pavement should be mitigated with site design. Conifer tree needles or leaf litter from deciduous trees can slow surface infiltration over time and may increase surface vacuum cleaning intervals. Other considerations include where plowed snow is piled. Snow can hold significant quantities of sediment and is best corralled onto non-paved areas to keep sediment from concentrating on paved areas after the snow melts.

- **Adjacent buildings**. In new and retrofit designs, removing existing impervious pavement and installing permeable pavement changes the soil moisture and movement of water. Nearby basements (typically within 3 m [10 ft]) will likely require waterproofing and drains, and foundations supporting slab-on-grade structures must have functioning footer drains.

7.3 Design

Once a site has been selected, some key elements to consider include:

- **One size does not fit all**. First and foremost, permeable pavement designs need to address owner needs and intended uses: Is the surface for cars? How many trucks will use it? Is it a pedestrian alley? Will it be used to reduce combined sewer overflows, control runoff volume, or achieve regulatory and/or water quality objectives? Pervious concrete may be a suitable choice for a residential road, while grass-planted grid pavement may be ideal for overflow parking. Some permeable interlocking concrete paver systems are suitable for pedestrian and vehicular areas but are not as suitable for bike paths due to their modular surfaces. Not all permeable pavement systems are equal, so consider the intended use and users. For example the pervious concrete

Figure 7-2
Pervious concrete basketball court provides infiltration and sound absorption for local residents, Neighborhood House Community Center, Seattle, WA
Source: SvR Design Company

basketball court shown in **Figure 7-2** provides a play surface and sound absorption for the adjacent residents living nearby.

- **Continue design coordination with other design disciplines**. Geotechnical engineers, civil engineers, architects, landscape architects, and other design professionals need to collaborate throughout the design process to ensure that design elements interact and function properly. For example, vegetation selection adjacent to the permeable pavements should be coordinated between the landscape architect, civil engineer, and maintenance staff to avoid aggressive plantings that could spread onto the permeable pavement.

- **Design for maintenance**. Maintenance is essential to the life cycle of permeable pavement and teams should establish a maintenance schedule in the design phase. This may be as simple as providing the project owner with a maintenance inspection checklist. More in-depth approaches might involve specifying vacuum equipment and settings, which can be obtained from a local pavement cleaning contractor or vacuum machine dealer and/or designs, which ease accessibility of maintenance vehicles.

- **Identify responsibilities and requirements**. In some jurisdictions, conventional asphalt or concrete public sidewalks are maintained by the adjacent property owners. When a sidewalk is converted to permeable pavement, it becomes not only a pedestrian pathway, but also a

Figure 7-3

Aggregate base/subbase for this porous asphalt pavement is clean and free of fines, Waste Management Woodinville, Woodinville, WA
Source: SvR Design Company

drainage facility and may fall under the management of a public utility or public works department. Similar circumstances have led some cities to change the way they manage newly installed permeable alleys. Many cities do not vacuum sweep alleys. When converted to permeable pavement, cleaning responsibilities for permeable surfaces will need to be decided among the appropriate city agency, and funding designated for the maintenance if not currently allocated. Budgets will need to be determined. In addition, various permeable surfaces may require different maintenance, so consideration of the associated equipment, staff training, and costs are important as well. A regenerative air vacuum machine can address bi-annual maintenance for most hard surfaces, but a leaf litter vacuum program applied more often may keep some permeable pavement surfaces clean for years. Watering and mowing will be required for grid pavements covered with turf grass as well as re-seeding as needed. Consider maintenance requirements early in the design phase, as well as what equipment will be available.

- **Public vs. private applications**. Permeable pavement suitable in private development is not necessarily acceptable in the public sector. Maintenance equipment, level of service expectations, and funding sources are different between the private and public sectors. This can affect the material and system selected. Specific stormwater design requirements such as limits on flow rates and management of targeted pollutants can also vary between public and private projects. These design requirements should be determined early in the process and should consider soil amendments, aggregate depths, overflow devices, and other design elements.

- **Base/subbase material selection**. Proper specification of base/subbase material is important for, not only the structural requirement of the paving system, but also for the infiltrative and temporary storage capacity of a system. If storage is required, open-graded bases/subbases with about 30% to 40% void space are recommended. In more arid regions, treatment is often the objective (rather than volume reduction) and base materials may have a lower percentage of void space, but still function adequately as a filter for stormwater passing through it (**Figure 7-3**).

One reason for infiltration failure in the past has been the use of specifications calling for dense-graded aggregate base material. Dense-graded base material used for non-pervious pavement has a high fines content. This is not suitable for permeable pavement applications since it does not allow for water to filter through the base/subbase and be temporarily stored before infiltrating into the native soils. Selection of open-graded aggregates should be investigated by visiting local supplier(s), who should be able to provide sieve analysis and possible porosity test results per ASTM or AASHTO standards upon request. Hardness and durability tests such as Los Angeles abrasion or magnesium sulfate soundness tests may be appropriate for aggregates used in permeable pavements subject to concentrated truck or bus traffic.

Figure 7-4
Permeable interlocking pavers serve as the drive access for multiple residences, Lyle Homes Development at High Point, Seattle, WA
Source: SvR Design Company

- **Availability of materials**. Regardless of the system, designers need to check with suppliers on the availability and character of specified materials. Specifications for gradation and petrography of aggregate materials will vary from region to region. A quarry in one area may not produce the same aggregates as a quarry in another part of the region. It is important to determine material availability early in the design phase. Most quarries that supply state-funded road projects will have aggregates suitable for use in permeable pavements, and such availability can help contain costs, especially if the quarry is close to the project site.

- **Update and use current specifications**. Compared to conventional pavements, permeable pavements are new. Industry recommended specifications are changing and improving. When starting a new project, request and review the latest available information. Call suppliers, manufacturers, installers, and other designers in the area to discuss historic and current installations. In addition, document any and all design and construction lessons learned for reference during the next permeable pavement project. Any new information forwarded to the ASCE Permeable Pavement Technical Committee may be reviewed and considered for updates in possible future editions of this report.

- **Not all projects have the same design intent**. Permeable pavements can be used for a number of applications, including stormwater treatment and conveyance, flow control, noise reduction, traffic calming, solar reflectance, water harvesting, context/historic sensitive design, and even as a thermal reservoir for ground source heat pumps supplementing nearby building energy needs (**Figure 7-4**). Specifications, related quality control, and quality assurance measures will vary to meet the design intent. For example, projects designed for water quality or flow control may have more quality assurance testing requirements than those used for non-drainage purposes such as for an alternative surface treatment to deter skateboarders from using a park path. Similarly, a street project intended for stormwater flow control will have stricter specifications on subgrade, base/subbase, and surface testing than a pavement installed only for pedestrian use. Not all projects are the same, and quality control, inspection, and testing for quality assurance needs to be matched to the project.

- **Overflow/back-up system**. Permeable pavements are designed to manage a range of storms, but not all of them. Therefore, larger storms must have a mechanism to pass larger flows through the system. These are typically designed as a bypass or overflow of the system and there are many different configurations available to consider (see **Chapter 1** for some examples). The pavement system design should always address the overflow path for the water that does not infiltrate into the soil subgrades. Overflow discharges should not result in erosion and/or the removal of sediments (and pollutants) at the overflow location and should not cause negative impacts to the property, adjacent properties, or downstream resources. Overflows are sometimes handled by conventional catch basins and sewer pipes, particularly in dense urban sites, and sometimes site conditions may allow for flows to be discharged to planted grass swales, infiltration ponds, or other non-critical vegetated areas. Most importantly, all designs must control overflows to prevent impacts to public safety or property or resource area damages.

7.4   Construction

Proper construction practices are key to the success of a permeable pavement system. Some important elements to consider include:

- **Qualified and experienced suppliers and contractors**. Permeable pavement requires different procedures for construction than conventional pavement. Therefore, qualified, experienced suppliers and installers are critical. Contractor qualifications should be outlined in the specifications and should require experience with projects of similar scope and size.

  Some industry associations provide installer training. While some training programs may only test knowledge and not skills, they make installers aware of the unique aspects of permeable pavement construction. This is a benefit to the project designer and owner. In addition, such programs require a minimum experience level (in years or project area) as a part of qualifying for earning recognition from the association. Project specifications should require that at least two to three people with the industry certification/experience be part of the installation crew for the permeable pavement.

  A training program is offered by the Interlocking Concrete Pavement Institute (ICPI) called the PICP (Permeable) Installer Technician Course for permeable interlocking concrete pavements (ICPI 2009). Concrete grid pavement installation training is covered in the ICPI Commercial Paver Technician Course (ICPI 2009). The plastic grid industry does not yet offer installation training.

  The National Concrete Ready Mix Association (NRMCA) offers a certification program for pervious concrete installers. This program has three levels: Technician, Installer, and Craftsman. The technician takes a course and must pass an exam. Besides these activities, the higher two levels also require varying levels of completed projects and work experience (NRMCA 2010). Additional information on NRMCA's Pervious Concrete Certification Program including a database of the certified technicians, installers, and craftsman is available on their website at www.nrmca.org.

  The porous asphalt industry provides general information on construction (NAPA 2008) and training webinars, but does not yet offer installer training with exam-based credentials. In lieu of certification, the installer should demonstrate past experience placing porous asphalt (recommend a minimum of three projects).

  Industry experience applies to those who install the surfaces and base/subbase. In some projects, the base installation can be subcontracted to a conventional pavement construction or excavation company. The subcontracting company should have a foreman on the job site with credentials indicating successful completion of permeable pavement industry training.

  In addition to having qualified installation crews, agencies, designers, and owners should perform product competence and quality tests by requiring evaluation of a test panel or mock up. The test panel should be installed by the same crew that will be completing the work on the project site. Once a test panel is evaluated and approved, it can be used as the minimum expected standard for the permanent application. For porous asphalt, it is not feasible to install a test panel given the small quantity of material typically required for a test panel (the NRMCA recommends a minimum 21 m² [225 sf] test panel for pervious concrete.) In such cases, a specifier can require that the installer provide at least three examples of porous asphalt that has been installed by the same crew for the owner's review. Even if the installer is certified, it is recommended to request examples/site visits of of previous projects within a certain radius of the project site. A range of 25 to 50 miles might be an option depending upon project location.

■ **Hold a pre-construction meeting.** For commercial and municipal projects, the specifications should include a request for a pre-construction meeting (Smith 2011) to discuss methods of accomplishing all phases of the construction operation, contingency planning, and standards of workmanship. The general contractor typically provides the meeting facility, meeting date and time. Representatives from the following entities should be present:

1. Contractor superintendent

2. Permeable pavement subcontractor foreman(s); responsible for excavation and placement of section

3. Permeable pavement manufacturer's representative

4. Testing laboratory(ies) representative(s)

5. Engineer or other owner's representative

The following items should be discussed and determined at the meeting:

1. Scope and schedule

2. Test panel (mock-up) location and dimensions

3. Plan for keeping subgrade protected from construction sediment and compaction

4. Plan methods for keeping all materials free from sediment during storage, placement and on completed areas

5. Plan for checking slopes, surface tolerances and elevations

6. Material delivery, timing, storage location(s) on the site, staging, paving start point(s) and direction(s)

7. Anticipated daily paving production and actual record

8. For PICP: diagrams of paving laying/layer pattern and joining layers as indicated on the drawings and monitoring/verifying paver dimensional tolerances in the manufacturing facility and on-site if the concrete paving units are mechanically installed

9. Testing intervals for sieve analyses of aggregates, as well as other material and quality assurance testing specific to each type of permeable pavement

10. Method(s) for documenting batches or unit paving packages delivered to the site

11. Testing lab location, test methods, report delivery, contents and timing

12. Engineer inspection intervals and procedures for correcting work that does not conform to the project specifications

Materials testing should be for the actual materials to be used on the site and not a representative sample from previous project sampling or an unidentified batch not used on the site.

Figure 7-5
Installation of this pervious concrete roadway takes careful planning, High Point Redevelopment Phase 1, Seattle, WA
Source: SvR Design Company

Figure 7-6
Sediment from unstabilized soils flowing to permeable pavements, Neighborhood House Community Center, Seattle, WA
Source: SvR Design Company

- **Protection of work area throughout construction**. The highest priority must be preventing and diverting sediment and stormwater from entering excavations for the permeable pavement base aggregates and pavement surface during construction (see **Figure 7-5**). Extra care must be applied to keeping sediment completely away from aggregates stored on site as well as from the permeable pavement surface. In some cases, it may be necessary to construct the permeable pavement before other soil-disturbing types of construction are completed. In this case, the permeable pavement must be completely protected for the duration of the construction activities. As shown in **Figure 7-6**, sediment from unstabilized landscape soils at this parking lot has clogged a portion of the pervious concrete panel. Stabilization of adjacent landscape areas and minimizing runoff from these areas will reduce the maintenance effort for permeable pavement systems

The options identified below will help prevent permeable pavement from becoming contaminated with sediment from construction activities and vehicles. One or more of these options should be decided upon in the project planning stages and expressed in the specifications and drawings.

1. Provide a staging and construction vehicle access plan. Establish temporary road or roads for site access that do not allow construction vehicle traffic to drive over and contaminate the permeable pavement base materials and/or surfaces with mud and sediment. Other trades on the jobsite need to be informed about using temporary road(s) and staying off the permeable pavement. The temporary road can be removed upon completion of construction and opening of the pavement surface to traffic.

2. Protect finished permeable pavement surface by covering it with materials such as a woven geotextile, tarp, or other material. The covering shall be weighted or anchored to keep it in place. Upon project completion and when adjacent soils are stabilized with vegetation, then the protective cover can be removed.

3. Construct the aggregate subbase and base, and protect the surface of the base aggregate with geotextile and an additional 50 mm (2 in.) thick layer of the same base aggregate over the geotextile. Thicken this layer at transitions to match elevations of adjacent pavement surfaces subject to vehicular traffic. A similar, more costly approach can be taken using a temporary asphalt wearing course rather than the additional base aggregate and geotextile. When construction traffic has ceased and adjacent soils are vegetated or stabilized with erosion control mats, remove geotextile and soiled aggregate (or the asphalt) and install the remainder of the permeable pavement per the project specifications.

Other practices such as keeping muddy construction equipment away from the permeable pavement, installing silt fences, staged excavation, and temporary drainage swales that divert runoff away from the area will make the difference between a pavement that infiltrates well or poorly. A more involved practice is a tire washing station for truck tires. Larger projects may require this level of cleanliness as trucks enter a muddy site. Truck tire washing equipment requires disposal of dewatered sediment. Truck tire washing stations should be located away from the site and not drain onto the permeable pavement and other infiltration facilities.

- **If changes are made that deviate from the specifications, obtain approval from designer before proceeding**. Many survey respondents noted that issues arose when materials were installed that did not meet the specifications, or a mix design was supplied that was not in accordance with the approved mix submittal. Proposed revisions or substitutions must be reviewed by the designer in order to confirm that the system would still meet the overall design intent and function. If revisions are proposed, it is vital that persons with an understanding of the implications provide guidance and input.

- **Minimize the compaction of the native subgrade**. In order to optimize the infiltration of the native undisturbed subgrade, it is critical to minimize the amount of compaction during preparation. This is a key difference from non-porous pavement applications where installers are instructed to compact and seal the subgrade prior to placement of the subbase. Options for minimizing compaction during construction include:

  1. Designate staging and construction traffic routes on the plans away from the permeable pavement and other infiltration facilities.

  2. During construction of the permeable pavement section, back out of the site during the excavation process, (i.e., as the operator excavates to grade the subbase material is placed immediately afterwards). Equipment is worked from the sides to avoid running vehicles over the subgrade and compacting it, affecting the permeability of the existing native soils.

- **Provide inspection checklists**. Construction oversight is enhanced by using inspection checklists to help organize and guide inspectors. Check with industry associations for system-specific construction checklists. Some items to include are: verifying installer and supplier have specified certificates/ certifications; reviewing the test panel; observing that erosion and sediment control and flow diversion measures are in place; verifying that the material/mix delivered to the site is the same material that was approved; providing quality assurance testing, such as an infiltration test (could be done on native subgrade and/or at the finish grade of the top wearing course); and verifying that erosion and sediment

protection measures are being adhered to throughout construction and after installation. It is also important that designers meet with inspectors to review design intent, particularly when permeable pavement systems are new to an agency or staff.

- **Keep subcontractors informed**. This is especially important on large projects that may last many months or years and involve several subcontractor trades. **Figure 7-7** shows that the protective cover over the pervious concrete sidewalk was removed by the landscape subcontractor during planting of adjacent areas; resulting in tracking of debris. As a result, the

Figure 7-7
Protective covering prematurely removed from pervious concrete sidewalk, resulting in clogging
Source: SvR Design Company

sidewalk had to be cleaned/vacuumed to restore permeability of the pervious concrete. All subcontractors need to be informed of the permeable pavement and its protection measures. This can reduce the risk of contamination and/or damage of the installed permeable pavement and related costs.

7.5 Operations and Maintenance

The following are general guidelines to follow for operations and maintenance of permeable pavement systems:

- **Safety, drainage and aesthetics.** Safety relates to the amount of maintenance required to ensure that the permeable pavement remains safe for pedestrian or vehicles. Drainage relates to the amount of maintenance required to ensure that the pavement surface is receiving and infiltrating and/or conveying flow as designed. Aesthetics means appearance acceptable to the owner. The frequency of the maintenance may be proportional to the severity of the problem and to the importance of the resolution. For example, the permeability of a porous asphalt or pervious concrete system clogged by the falling leaves can often be restored with a regenerative vacuum sweeper. Weeds growing in a permeable gravel parking lot, which may be an eyesore to an owner and yet still meet the drainage requirements, can be removed manually or by other non-chemical means, if needed.

- **Stabilization of adjacent areas**. The frequency of maintenance to remove sediment deposits from a permeable pavement depends upon continued stabilization of adjacent areas that flow onto the surface. Take a proactive approach by addressing the source of the problem, such as maintaining landscape areas and not having exposed soil.

- **Tracking locations in agency/homeowner association records**. Utility service crews who may install service lines or excavate for utility maintenance in the future should be aware of permeable pavements to avoid potential contamination or improper restoration. Some public agencies have begun tracking information in their GIS or on other maps so the pavement can be identified when developers or service prodivers apply for a permit. As-built records should be available.

- **Create a maintenance schedule and inspection checklists**. Prior to close-out of a project, it is important that owners have an operations and maintenance plan in place. The designer should provide general guidelines for the frequency of maintenance tasks and equipment needs. General maintenance guidelines for owners have already been developed by some agencies, many of which can be used as a reference for developing site specific maintenance plans (see **References**). General recommended practices for operations and maintenance are in **Chapter 1 and Chapter 8**, and details for specific pavement types are included in **Chapters 2 through 6**. A maintenance inspection checklist, such as the one included in **Chapter 8**, is another way to help guide maintenance crews on key observations to help identify maintenance needs.

Figure 7-8

Educational signage at the Humboldt Penguin Exhibit. This exhibit sign educates the public on the design and environmental benefits of pervious concrete pavement, Woodland Park Zoo, Seattle, WA

Source: SvR Design Company

- **Inform users of the permeable pavement facility by signage**. Until permeable pavement systems become more widely recognized by the public, agencies should consider signage to inform users of how the systems operate and their intended function. The signage can also be useful to maintenance crews for informing them of the presence of the systems and what to avoid in terms of impact on the pavement function and performance (**Figure 7-8 and Figure 8-21**).

7.6 Summary

For a successful permeable pavement project, collaborate and coordinate with the design team and owners throughout all levels of site design, site development, and project construction. Choose materials and locations that consider site elements, adjacent property conditions, intended users, and system design goals. Employ trained and experienced contractors and installers. Inform crews, inspectors, owners, and users about the presence and purpose of installed systems. Maintain installed systems to provide functionality and aesthetics.

7.7 References

American Concrete Institute (ACI) (2013). "Specification for Pervious Concrete Pavement." *ACI 522.1-13*, American Concrete Institute, Farmington Hills, MI.

Interlocking Concrete Pavement Institute (ICPI) (2009). *PICP Installer Technician Certificate Course*, student manual, 1st Ed., Interlocking Concrete Pavement Institute, Herndon, VA.

Kevern, J. T., Schaefer, V. R., Wang K. (2009). "Design of Pervious Concrete Mixtures." Department of Civil and Mechanical Engineering, University of Missouri, Department of Civil, Construction and Environmental Engineering, Iowa State University.

National Asphalt Pavement Association (NAPA). Hansen, K. (2008). "Porous Asphalt Pavements for Stormwater Management; Design, Construction and Maintenance Guide." National Asphalt Pavement Association.

National Ready Mixed Concrete Association (NRMCA) (2010). *Text Reference for Pervious Concrete Contractor Certification*, Publication 2PPCRT, Silver Spring, MD.

Environmental and Water Resources Institute (EWRI) (2010). "Pervious/Porous Installation Experience for Development." *Permeable Pavements EWRI Report: Avoiding Failures Questionnaire*, Survey Monkey Internet Accessed (2010).

Smith, D. R. (2011). *Permeable Interlocking Concrete Pavement*. 4th Ed., Interlocking Concrete Pavement Institute (ICPI), Herndon, VA.

8 Maintenance

8.1 Introduction

While site selection and proper construction are the most critical factors in establishing that permeable pavements function as designed, all must be inspected and maintained to maximize performance and longevity. In order to maintain permeability, permeable pavements require vacuum sweeping and designs/management that prevent the accumulation of fines and sediments. Maintenance requirements for permeable pavements are not excessive and do not inhibit their use for the following reasons:

- **Limited sweeping frequency**—Typical minimum maintenance is vacuum sweeping two times per year. This may be no more prescriptive than standard pavement sweeping practices. The cost for biannual vacuum sweeping will typically be less expensive than the typical maintenance required for rain gardens/bioretention areas treating the same area.

- **Efficient sweeping methods**—Vacuum sweeping methods are more efficient and effective at removing large sediments and fines that can cause stormwater pollution, buildup in stormwater conveyance system and/or BMPs, and increased stormwater system maintenance costs. These methods are becoming cost-competitive as contractors and municipalities replace outdated sweepers with vacuum sweepers. More maintenance contracts are requiring updated equipment in order to increase efficiency, lower costs, and/or meet stormwater runoff quality goals.

- **Increased sweeping requirements for all pavements**—Street cleaning is becoming an increasingly typical maintenance activity for all types of pavements on public and private properties. The recommended maintenance frequency of two times per year may coincide with local guidelines or regulations. Street sweeping requirements may be included in regulatory stormwater permits, such as the EPA NPDES Municipal Stormwater Permit and/or under local bylaws/ordinances, and is therefore not realized as any additional cost.

- **Cost savings**—Permeable pavements typically result in the reduction or absence of catch basins and pipes for stormwater management. This results in reduced or eliminated inspection costs and disposal costs for catch basins, manholes, and pipe cleaning operations.

In northern climates, cost savings are further realized by eliminating sanding for permeable pavements. Spring street and catch basin cleaning focuses on sand removal that accumulated during the winter season. Reduced or eliminated sand applications results in reduced total sweeping volumes, disposal costs, and potential for unintended discharge of sand into water bodies from rainfall events that occur prior to spring sweeping schedules.

- **Pollutant load reduction credit**—Pollutant load reduction credits due to street sweeping may be applicable in some jurisdictions and may help in achieving total maximum daily loads (TMDL) reductions, even if the permeable pavement is not given such a credit.

8.2   Clogging Sources and Prevention Methods

A common permeable pavement performance concern is the clogging of the surface and subsurface void spaces. Clogging can occur when sediment and/or vegetative litter/organic materials are transported onto the pavement surface or are deposited as runoff from adjacent land area flows onto the pavement (**Figure 8-1**).

Sediment

Clogging can be of a greater concern with areas of exposed fine soils and in areas with unstabilized soils, such as construction sites or eroding areas. Sediment laden flows onto pavement during construction from unstabilized sites are a common concern (**Figure 8-2**).

Figure 8-1
Porous asphalt pavement clogged by runoff sediments from an unstabilized adjacent area, Philadelphia, PA
Source: North Carolina State University (NCSU)

Impacts from Adjacent Soil Types

Bean et al. (2007) showed that even when maintenance (street sweeping) has been neglected for up to 20 years, permeable pavements can still be permeable, if the surrounding soils that were transported onto the pavement are permeable. Several tests of older permeable pavements located along the North Carolina barrier islands and in the state's sand hill region demonstrated infiltration rates usually exceeding 13 mm/hr (0.5 in./hr). This is because the pavement was clogged by sand, rather than silt or clay particles. Designers and regulators should consider acceptable infiltration rates and adjust surface cleaning schedules as needed. For North Carolina, a 30-year analysis of rainfall intensities indicated that less than 2% (as measured hourly) exceeded 13 mm/hr (0.5 in./hr).

Vegetation

While trees provide shading to cool pavement temperatures, increase property values, and improve aesthetics, consideration is required on species selection and leaves that might clog the surface. Trees often border parking lots or grow in islands within lots. In either case, leaf deposition on limited areas of adjacent permeable pavement may present temporary and acceptable reduction in surface infiltration (**Figure 8-3 and 8-4**). Even with larger trees, there may not be a significant concern, depending on the species and parking lot surface cleaning practices. Grass clippings and tree litter should not be blown on to pavement. Sweeping practices may reduce or eliminate the concern for clogging impacts. The pavement sweeping schedule should be planned accordingly, while considering seasonal frozen conditions, including conditions in northern climates.

Materials Storage/Stockpiling

Sand, soil, mulch, or other landscaping materials are often temporarily stored on pavement during seasonal maintenance activities and can be a clogging source. This threat can be eliminated by simply

Figure 8-2
Nearby construction increases the risk of
sedimentation of permeable pavement
Source: NCSU

Figure 8-3
Leaf and tree detritus collects in PICP openings
located underneath a tree line, Chicago, Illinois
Source: NCSU

Figure 8-4
Organic material deposition in PICP openings
from nearby trees
Source: NCSU

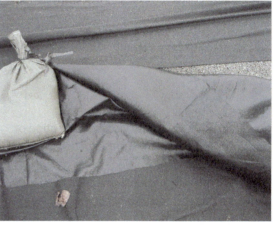

Figure 8-5
Tarp being used to protect pervious concrete
from clogging, Seattle, WA
Source: E. Fassman

placing a tarp beneath and on top of the stored material or by moving the material to impervious
pavement. **Figure 8-5** shows a tarp being used to temporarily protect a pervious concrete sidewalk from
clogging during construction of a street project. While this side of the street was complete with stabilized
soils and planted grass, a large soil stockpile was across the street. The sidewalk was protected from
windblown sediment and potential sediment run-on from ongoing adjacent construction activities.

Snow storage piles often contain accumulated sediment and debris that can clog voids. Piling snow on
permeable pavement is not recommended and should be prohibited wherever possible. If this guideline is
not followed, post-winter surface cleaning will be required to maintain permeability.

8.3 Primary Inspection and Maintenance Tasks

Permeable pavement maintenance can be less costly when compared to other stormwater management practices. **Table 8-1** provides conservative, planning-level maintenance cost comparisons for the Philadelphia, Pennsylvania, area. Costs are based on controlling 25 mm (1 in.) rainfall coming from impervious areas and draining into the practice for infiltration and/or slow release.

Table 8-1 Typical Maintenance Cost Comparisons for Various Stormwater Management Practices in 2009

MANAGEMENT TOOL	ANNUAL OPERATION & MAINTENANCE COSTS ($/IMPERVIOUS ACRE/YEAR)
Permeable pavement	$2,400
Subsurface infiltration	$2,900
Green roof	$4,000
Bioretention	$3,100
Street tree	$1,800

Source: Data from Vanaskie et al. 2010

Inspection

Inspections should include visual examination of the permeable pavement surface for accumulated dust, sediment, debris and vegetation growth. **Figure 8-6** shows the visual inspection of a pervious concrete lot where sediment and leaf debris has accumulated.

Ponding areas observed during or immediately following a rainfall event can indicate clogged areas, but may also be caused by inadequate subsurface drainage. Emptying a bucket or large bottle of water on the pavement surface and observing the length of time required for the water to infiltrate or how large the wetted area spreads can indicate pavement clogging. As shown in **Figure 8-7**, surface clogging is indicated on the lot by emptying approximately 4 liters (1 gal) of water on to the pavement, which runs off rather than infiltrating.

Figure 8-6
Visual inspection of a pervious concrete lot shows sediment and leaf debris accumulation
Source: NCSU

A systematic assessment of new and in-service pavement surface infiltration rates can be done using ASTM C1701 *Standard Test Method for Infiltration Rate of In Place Pervious Concrete* and C1781 *Standard Test Method for Surface Infiltration Rate of Permeable Unit Pavement Systems*. Both are rapid and inexpensive tests with comparable results. While developed for pervious concrete, the C1701 *Standard Test Method* has also

Figure 8-7
Demonstration of clogged pervious concrete surface using one gallon of water, Savannah, Georgia
Source: NCSU

Figure 8-8
Inexpensive test apparatus for implementing ASTM C1701 and C1781 methods for assessing surface infiltration rates and cleaning needs for permeable pavements (PICP shown here)
Source: Interlocking Concrete Pavement Institute (ICPI)

been successfully used to measure the surface infiltration rate of other pavements, including PICP, porous asphalt and grid pavements (Smith et. al 2012; Bean et al. 2007). Similarly, reports using a modified version of C1701 to test PICP, pervious concrete, and porous asphalt as part of an evaluation of these pavements by the US EPA have been completed (Borst et al. 2010). **Figure 8-8** shows the inexpensive materials and ease of deployment for the test apparatus and methods described in ASTM C1701 and C1781.

Regarding acceptance criteria for new and in-service pavements, the Interlocking Concrete Pavement Institute and the National Concrete Ready Mix Association recommend a minimum 250 cm/hr (100 in./hr) surface infiltration rate for new pavements and a minimum of 25 cm/hr (10 in./hr) for in-service pavements. Similar values are recommended for porous asphalt. However, new sand-filled grid pavements typically have a lower recommended minimum rate of 64 cm/hr (25 in./hr) due to smaller voids. These new and in-service recommendations are based on professional experience and account for uncertainty from variable clogging of the surface, point measurements for area evaluation, infrequent inspections, a factor of safety, and decreasing surface infiltration rates before subsequent measurement, while accommodating most rainfall rates and run-on from adjacent impervious areas.

Inspection and maintenance tasks common to all permeable pavements are as follows:

1. Measure surface infiltration rate (i.e., clogging has not occurred)

2. Preventive (regenerative) regular surface sweeping to maximize performance and longevity

3. Weed removal as required on grid pavement, plus mowing and watering if grassed. Weed removal from PICP and non-grid pavements can indicate sediment deposition and a need for surface vacuuming.

4. Restorative vacuum sweeping may be required if the pavement is severely clogged or if regenerative sweeping does not adequately restore infiltration.

5. If the permeable pavement has underdrains, outlet(s) should be checked to verify free flow of water.

6. Stain removal as necessary

7. Routine inspections for wear, particularly at edges of pavement and in areas of heavy use

Maintenance

Surface Cleaning

From a permeability standpoint, surface cleaning is the most important maintenance task for maintaining long-term functioning and viability. Vacuum sweeping can mechanically loosen surface sediments and material via contact by bristles with the pavement and remove clogging material from the surface via suction. Vacuum sweeping can serve as a preventive and restorative measure.

Preventative Surface Cleaning

Regular preventive surface cleaning with the appropriate street sweeping equipment reduces the likelihood of surface clogging. Two cleanings per year are sufficient for most permeable pavement applications if other preventive measures in **Section 8.3 Primary Inspection Maintenance Tasks** are followed. Spot vacuum treatments may be employed in areas of localized clogging. A pavement sweeping schedule should be prepared for the specific property and pavement use.

Rehabilitative Surface Cleaning

If the permeable pavement has been neglected for several years or excess sedimentation has clogged the surface, preventive cleaning may not be sufficient to restore surface infiltration. Therefore, restorative vacuum cleaning may be required to remove clogging sediments and restore a pavement's infiltration rate. The type of equipment needed for preventive cleaning or restorative cleaning are different.

Regenerative maintenance performed at the recommended interval (twice per year) should be sufficient to maintain infiltration rates above the recommended minimum infiltration rates. However, when regenerative maintenance is not performed over extended periods, sediment accumulation and compaction into voids typically requires removal via restorative maintenance.

Types of Pavement Cleaning Equipment

Street cleaning equipment perform similar functions, but differences in operation make them more appropriate for certain situations. There are three types of pavement (street) cleaning equipment: mechanical, regenerative air, and vacuum.

Mechanical street sweepers are currently the most common street sweeper in use on impervious pavements (**Figure 8-9**) and their primary purpose is to remove surface litter. Mechanical street sweepers rely on brush sweeping and collection of sediment from the surface, which does not use vacuum force. Brushes initially move litter and sediment to the machine's center, and another brush then lifts them into a conveyor belt to the machine's hopper. The brush bristles can penetrate some types of permeable pavements, but may not collect fine dust particles. Mechanical sweepers are available in a range of sizes for use on sidewalks, parking lots and streets.

Figure 8-9
Mechanical street sweeper
Source: Elgin Sweeper Company

Regenerative air street cleaners are the second most common street cleaner (**Figure 8-10**). Regenerative air sweepers have a rotary sweeper that loosens dirt and sediment, which can then be vacuumed into the hopper. The rotary brush ahead of the box and vacuum suction loosen dust and sediment near the pavement surface. Fast moving air is recirculated within a box mounted across the entire truck width, which creates a vacuum directly over the pavement surface. The vacuum force can be adjusted by controlling the machine speed. This machine can be used at regular intervals for preventative surface cleaning to remove loose sediment accumulation.

Figure 8-10
Regenerative air sweepers
Source: Elgin Sweeper Company

Vacuum sweepers are typically the most expensive type of street sweepers, but often can be equipped for multiple functions beyond traditional street sweeping (**Figure 8-11**). They have models that can come equipped with a hose attachment that may be used for a variety of tasks, including cleaning catch basins, manholes, or pipes. The machine is equipped with rotary brushes for loosening surface dirt prior to vacuuming. A strong vacuum is applied to an intake nozzle positioned on the pavement that lifts particles from the surface and within the pavement. The intake is approximately one (1) meter (3 ft) wide and applies

Figure 8-11
Vacuum sweeper
Source: Elgin Sweeper Company

a vacuum force at least twice that of a regenerative air sweeper. As a result, vacuum sweepers can be used for restoration of clogged permeable pavement surfaces. However, the narrower width of suction can increase cleaning duration.

The ability of this machine type to remove sediment varies with the pavement type, the depth of the sediment, and moisture content. A study by Chopra et al. (2010) demonstrated variable restorative effects of this machine type on the surface infiltration rates of Flexipave™, porous asphalt, pervious concrete, and two types of PICP. The authors concluded that vacuum machines were effective in restoring these pavements when severely clogged, albeit with varying recovery rates for surface infiltration.

Vacuum sweepers on PICP may be calibrated to adjust the force of the vacuum for the specific application. Too powerful a force may remove aggregates within PICP beyond the depth desired. Hunt (2011) demonstrated the ability of this machine type to remove as much as 75 mm (3 in.) of aggregates soiled with sediment from within PICP openings. Results for a PICP parking lot surface with no cleaning for seven years are shown in **Figures 8-12 and 8-13**. **Figure 8-14** shows a schumtzdecke, or "dirty roof," of sediment formed by trapped sediment at the surface. After applying vacuum equipment to remove soiled aggregate and sediment, the pavement continues trapping sediment at the surface. Vacuuming is most effective when the sediment within PICP and other permeable pavement surfaces is as dry as possible.

Figure 8-12
Before vacuum cleaning of a PICP parking lot
surface that was not cleaned for seven years
Source: ICPI

Figure 8-13
After vacuum cleaning of a PICP parking lot surface
Source: ICPI

Figure 8-14
PICP opening showing trapped sediment at the
top with aggregate under it
Source: NCSU

Figure 8-15
PICP openings and joints must be filled with
aggregate to prevent tripping hazards.
Source: NCSU

Figure 8-16
Replacement is simple and done with a broom or
with mechanical sweepers for larger areas.
Source: NCSU

Figure 8-17
Hollowed out concrete grid pavement voids
penetrated by mechanical sweeper bristles,
leaving loose sand from large voids on pavement
surface to be swept back into voids.
Source: NCSU

Vacuum sweepers are powerful and can withdraw aggregate from between pavers. They can remove aggregate from PICP openings and joints. Therefore, PICP surface cleaning will require aggregate refilling to the bottom of the paver chamfers (**Figure 8-15 and 8-16**). If openings or joints remain unfilled, they can pose a tripping hazard. In select cases, aggregate color may be important to the project owner and should be confirmed in such cases. Refilling openings or joints is a fast and easy process that is usually achieved by manual or mechanical broom sweeping of aggregate into openings and joints. Restorative surface cleaning with vacuum equipment and replacement of fresh aggregates is estimated in 2012 at $2,700/ha ($8,000/ac). Preventative surface cleaning is approximately $330/ha ($1,000/ac). The cost difference emphasizes the savings from regular cleaning versus deferred maintenance.

The extent of aggregate depth to be removed depends on the penetration depth of sediment. The depth can be determined by withdrawing aggregate from openings or voids with a tool until clean stone appears.

Machine operator care must be taken to adjust the vacuum engine speed and resulting suction to limit aggregate removal. The extent of aggregate removal can be quickly assessed by making a test pass with the vacuum equipment and inserting a pencil into the vacuumed joints to verify the depth of removed aggregate. The speed of the vacuum machine is then adjusted as needed.

Vancura et al. (2012) evaluated cleaning on three pervious concrete sites: an alley apron, alley, and low-volume residential street. The ages were one year, two years, and one month, respectively, and each had varying degrees of clogging. Cleaning was done using either a regenerative air cleaning machine, an flexible 200 mm (8 in) diameter hose attached to a vacuum machine, or a vacuum cleaning equipment. Rather than measure surface infiltration using ASTM C1701 or other test method, photomicrographs of extracted cores were taken before and after to evaluate cleaning effectiveness within the pavement cross sections. Soil particles generally did not penetrate further than 13 mm (0.5 in.) into the pavement surface.

All cleaning methods were able to remove sediment down to 3 mm (0.12 in.) below the surface, but none removed sediment deeper than 3 mm (0.12 in.). The number of passes made with the equipment was not indicated. Two of the three surfaces maintained observed permeability after cleaning. Vancura et al. (2012) concludes that consolidation of clogging material dictates permeability, rather than quantity of clogging material. Material left in the voids may not have consolidated, and therefore allowed water passage after cleaning. Their work indicates that surface clogging rate depends on location, sediment sources, antecedent, and regular surface vacuuming of organic material from nearby vegetation and eroding soil.

Research by Mata (2008) has shown that sand collected on a pervious concrete pavement will not penetrate the pavement more than 25 mm (1 in.) below the surface. While pervious asphalt was not tested, it is assumed results would be similar. However, when a fine fraction of soil, silts, or clays was a part of the applied sediment, the clogged

Figure 8-18
Pavement types with small openings or gaps (such as PICP) will not be easily reached by sweeper bristles.
Source: Elgin Sweeper Company

zone was much deeper, often at or near the bottom of the pervious concrete section. Restorative maintenance of a porous asphalt or pervious concrete system clogged by fines appears to be difficult, especially if clogging occurs deep in the pavement profile. This condition emphasizes the need for regular preventive maintenance with a regenerative air street sweeper to maintain surface infiltration rates.

Concrete grid pavement surfaces can be cleaned when filled with open-graded aggregates or sand. In such cases, concrete grid pavement trap most fine sediment at the surface. For grid openings filled with sand, the thickness of the "schmutzdecke" ranges from 6 to 13 mm (0.25 to 0.5 in.) (James and Gerrits 2003; Bean et al. 2007). Sweeper bristles on a mechanical sweeper can penetrate the larger openings in grid pavements to help break apart and remove this layer (**Figure 8-17 and 8-18**).

Table 8-2 presents clogging depths of permeable pavements and recommended equipment for preventing or restoring clogged surfaces. All pavement types can maintain adequate surface infiltration rates if preventive sweeper maintenance is regularly conducted. If pavement cleaning is neglected, clogging can occur and at depths that can impair infiltration.

Table 8-2 Street Sweeper Assignment to Permeable Pavements

PAVEMENT TYPE	SCHMUTZDECKE DEPTH	PREVENTIVE SWEEPER	RESTORATIVE SWEEPER
Concrete grid pavement filled with sand	13 mm (0.5 in.)	Mechanical	Mechanical
PICP filled with aggregate	25 to 50 mm (1 to 2 in.)	Regenerative air	Vacuum
Pervious concrete & porous asphalt	Sand >13 mm (0.5 in.) Silt/Clay >6 mm (0.25 in.)	Regenerative air	Vacuum

Source: Data from Vanaskie et al. 2010

Winter Maintenance

Proper winter maintenance is critical for continued functionality of permeable pavements in northern climates. The following minimum practices are recommended.

- Prohibit sand application whenever possible.
- Prohibit stockpiling/storage of snow plowed from other areas and post "no sanding" signs.
- Inform and train operators (municipal/private maintenance staff).
- Clean up sand. If sand is applied, it must be removed as soon as possible with vacuum sweeping equipment.

Studies show permeable pavements can reduce deicing material use by as much as 75% compared to conventional pavements (Houle et al. 2008; Roseen et al. 2012). The pavement surface is warmed by solar radiation. The surface voids, base, and soil moisture significantly delay frost penetration into the underlying soil. The presence of surface snowmelt water is often sufficient to thaw the underlying ice and soil, thereby allowing infiltration. Surface drainage through permeable pavements will not allow puddles or a brine of deicing materials to accumulate on the surface.

Surface infiltrations reduces the potential for surface icing and thereby reduces the need for deicing chemicals. However, deicers may be applied to reduce ice or snow accumulation in freezing temperatures especially when sunlight is not effective or cannot reach surfaces to assist in snow/ice melting. Deicing materials should be used sparingly or not at all during the first service year of pervious concrete and are not recommended for long term use.

Removing Unwanted Vegetation

Vegetation in permeable pavements can indicate excess sediment accumulation since the sediment provides a media for vegetation to survive. It is imperative to prohibit unwanted vegetation. Weeds should be removed manually whenever possible. Herbicides should be avoided due to potential contamination of the subsurface, groundwater, or surface waters via underdrains. As weeds mature, the biomass becomes harder to remove and removal may damage the pavement structure. For pavements designed to receive grass, mowing will be required. Catch grass clippings before they are deposited on the pavement, since the accumulation of grass thatch can further limit infiltration.

Removing Oil and Grease Stains

The designer has the option of selecting pavement types that hide stains better than others. Porous asphalt does not show stains as much as other pavement types due to its darker color. However, the same pavement types that hide stains well also have low albedo, which might conflict with LEED® light reflectivity goals. Many permeable pavements are installed as parking areas and the owner may allow the pavements to stain from normal use. If stains cannot be tolerated, then a biodegradable detergent should be considered for cleaning. For PICP, detergents can be used to help lessen visible stains. They are applied to the stain and allowed to "soak." The following day, the stain remover is pressure washed from the pavers (**Figure 8-19 and 8-20**). The majority of stain is typically removed but a faint mark may remain. Permeable pavers may also be replaced completely if the stains are considered unacceptable after cleaning.

Figure 8-19
A stain remover applied to PICP is Ideally allowed to set overnight.
Source: NCSU

Figure 8-20
Subsequent pressure washing of PICP after the stain remover has set
Source: NCSU

Signage

One or more signs should be placed in a visible location identifying the pavement as permeable and indicating prohibitions. This information should also be provided to all service contractors for the property and as appropriate included in their contract agreements that may engage in any of these activities including landscaping, snow plowing, or facilities related activities that effect use, maintenance, or storage on or near the pavement. **Figure 8-21** shows some possible language that may used on a sign to indicate prohibited activities for a permeable pavement application. The specific prohibitions or recommended maintenance information would be specific to the pavement design and use. A sign should be placed in a location that may be observed by the maintenance personnel responsible for these activities.

Figure 8-21
Recommended signage for maintenance staff
Courtesy NCDENR

Table 8-3 provides a summary of recommended minimum permeable pavement maintenance activities and frequency.

Table 8-3 Permeable Pavement Maintenance Summary

TASK	FREQUENCY	NOTES AND DRIVERS
Inspection	Annually	Ensure pavement is draining effectively and no sediment accumulation on surface
Street sweeping (preventive)	2 to 4 times a year	Using a regenerative air street sweeper for most pavements is appropriate
Stain removal	Per client desires	Stains may be tolerated. If stain must be removed, consider a biodegradable detergent.
Weed removal	Do upon inspection	Remove manually whenever possible.
Mowing	In "growing season" 1 to 2 times per month	Applies only to vegetated permeable pavements
Plowing	As needed	Avoid scraping of pavement while plowing

8.4   Checklists 3 and 4

Detailed checklists, **Checklist 3: Annual Permeable Pavements Inspection** and **Checklist 4: Annual Permeable Pavements Maintenance**, are provided on the following pages.

CHECKLIST 3 ☑

Annual Permeable Pavements Inspection

☐ **1. RUN-ON/TRACK-IN**

Check all areas around permeable pavement perimeter for signs of sediment run-on or sediments tracked in from adjacent properties or roads.

☐ **Mitigation action**: Implement mitigation methods appropriate to reduce runoff or tracked in sediments.

☐ **2. VEGETATION/ SOILS/SLOPES**

Inspect vegetation around permeable pavement perimeter for coverage, quality and soil stability, especially areas that slope towards the pavement.

☐ **Mitigation action**: Stabilize all soils and replant grass or groundcover as needed. Ground covers are typically preferred over mulch. Replace dead or unhealthy vegetation intended for stabilization. Incorporate compost amendments into soils that will be re-vegetated to assist with moisture retention, germination and/or long-term health. If possible, relocate soil stockpiles to impervious pavements. Otherwise, cover stockpile and place barrier between permeable pavements and stockpiles.

Inspect pavement for presence of unintended vegetation, which can affect infiltration and indicate excess sediment accumulation in voids.

☐ **Mitigation action**: Vegetation should be removed manually. Avoid use of herbicides, which can be transported into unintended environments (e.g., subsoil, groundwater or surface waters via discharge).

Inspect grid pavements for grass coverage, soil or aggregate erosion, scour and unwanted growth.

☐ **Mitigation action**: Replenish lost soil or aggregate. Re-seed bare soil areas as needed. Fertilize grass in the spring and fall. Limit applications if runoff may carry nutrients into nearby waters.

☐ **3. ORGANIC MATTER BUILD UP**

Inspect for any excessive build up of organic materials or other debris (fallen tree leaves, mulch from adjacent areas, etc.)

☐ **Mitigation action**: Sweep up affected areas and consider alternative methods of vegetated cover/ mulching, or adjust slopes to reduce pavement buildup.

☐ **4. UNDERDRAINS/OUTFALLS/MONITORING WELLS/CLEAN OUTS**

Check that underdrains, outfalls and other flow paths allow for unobstructed water flow . Check monitoring wells (inspection ports) for standing water levels following a major storm.

☐ **Mitigation action**: Address clogged pipes via cleanouts. Clean outfalls and remove any obstructions. Recheck monitoring wells for adequate storage recovery.

*This information is a suggested framework for a checklist and should not be considered an exhaustive or complete list of all items that should be reviewed. Advice and direction from a competent professional in the field should be sought for site specific application of any and all material included in this report.

Source: © VHB

CHECKLIST 4 ☑

Annual Permeable Pavements Maintenance

☐ **1. VACUUM SWEEPING: 2 TIMES ANNUALLY (SPRING/FALL)**

☐ a. Vacuum surface with regenerative air equipment, adjust vacuuming schedule per sediment loading and/or sand deposits (e.g., following winter maintenance).

☐ **2. WINTER MAINTENANCE**

☐ a. Remove snow with standard plow or snow blowing equipment; monitor surface ice. Sand should not be applied to permeable pavements due to subsequent surface clogging. Applying deicing materials to permeable pavements should be either avoided or done so on a limited basis since studies have shown that permeable pavements require 75% less salt than conventional pavements per season and they can infiltrate and accumulate in the soil subgrade or be discharged via underdrains.

☐ b. **Permeable interlocking concrete pavement (PICP)**—If traction is required, ASTM No. 8 stone (or similar jointing material) should be used.

☐ c. **Pervious concrete**—Deicing materials can damage the cement in the pervious concrete mixture and result in disintegration or spalling of the surface, particularly if used over a newer application. Deicers should never be applied within the first year of pervious concrete installation but may be used on a limited basis on older installations.

☐ **3. TESTING FOR PERMEABILITY AND REHABILITATION FOR PERMEABLE PAVEMENTS**

☐ a. **Surface pavements suspected of having decreased permeability**—Test surface infiltration rate using ASTM C1701 or C 1781 for minimum acceptable rate (typically 250 mm/hr [10 in./hr]). Perform preventative or restorative maintenance when water ponds for more than 30 minutes after a storm.

☐ b. **Cleaning for rehabilitation of permeability**—Vacuum surface with restorative vacuum sweeper (more powerful than regenerative air equipment). to remove clogging sediments and material. Permeable pavements should not be washed with high-pressure water systems or compressed air units, unless part of a specific rehabilitation process.

☐ c. **PICP permeability rehabilitation**—Refill joints with clean aggregate, sweep surface clean.

☐ d. **Permeability after rehabilitation for all Permeable pavements**—Test infiltration rate again per ASTM C1701 with a minimally acceptable rate of 250 mm/hr (10 in./hr).

☐ **4. REPAIRS TO CRACKS OR DEPRESSIONS**

☐ a. **Porous asphalt and pervious concrete damage**—It is best to replace damaged pervious concrete areas with a visually and functionally similar pervious concrete mixture. Very thin repairs are inadequate and full-depth patches are required in most situations. Porous asphalt and pervious concrete can also be repaired with conventional (impervious) materials in selected areas raveled or cracked with depressions greater than 13 mm (0.5 in.) deep if pervious materials are not available. Do not cover more than 10% of the porous asphalt or pervious concrete pavement area with conventional paving materials.

CHECKLIST 4 ☑

Annual Permeable Pavements Maintenance *(continued)*

4. REPAIRS TO CRACKS OR DEPRESSIONS *(continued)*

☐   **b.**   **PICP damage**—Replace cracked pavers and refresh jointing aggregate; replenish aggregate in joints if aggregate is missing more than 13 mm (0.5 in.) from chamfer bottoms on paver surfaces.

☐   **c.**   **Grid pavement**—Replace pavers if cracking diminishes structural or hydraulic performance. Replenish aggregate if rutting or settling leaves aggregate more than 13 mm (0.5 in.)below pavement surface.

*This information is a suggested framework for a checklist and should not be considered an exhaustive or complete list of all items that should be reviewed. Advice and direction from a competent professional in the field should be sought for site specific application of any and all material included in this report.

8.5 References

Bean, E. Z., Hunt, W. F., and Bidelspach, D. A. (2007). "Field Survey of Permeable Pavement Surface Infiltration Rates." *Journal of Irrigation and Drainage Engineering*, 133(3), 247–255.

Borst, M., Rowe, A. A., Stander, E. K., and O'Connor, T. P. (2010). "Surface Infiltration Rates of Permeable Surfaces: Six Month Update (November 2009 through April 2010)." EPA/600/R-10/083, United States Environmental Protection Agency (US EPA), National Risk Management Research Laboratory, Edison, NJ.

Chopra, M. B., Stuart, E., and Wanielista, M. P. (2010), "Pervious Pavement Systems in Florida—Research Results." *Low Impact Development 2010: Redefining Water in the City 2010*, American Society of Civil Engineers (ASCE), Reston, VA.

Houle, K. M., Ballestero, T. P., Roseen, R. M., and Heath, D. (2008). "Winter Performance Assessment of Permeable Pavements: A Comparative Study of Porous Asphalt, Pervious Concrete, and Conventional Asphalt in a Northern Climate." Master of Science, University of New Hampshire, Durham, NH.

Hunt, W. F. (2011). "Urban Waterways—Maintaining Permeable Pavements." North Carolina Cooperative Extension, College of Agriculture and Life Sciences, North Carolina State University, Raleigh, NC.

James, W., and Gerrtis, C. (2003). "Maintenance of Infiltration in Modular Interlocking Concrete Pavers with External Drainage Cells." *Practical Modeling of Urban Water Systems*, Monograph 11, W. James, ed., Computational Hydraulics International, Guelph, Ontario, 417–435.

Mata, L. A. (2008). "Sedimentation of Pervious Concrete Pavement Systems." *Ph.D. Dissertation*, North Carolina State University, Raleigh, NC.

Roseen, R. M., Ballestero, T. P., Houle, K. M., Heath, D., and Houle, J. J. (2012). "Assessment of Winter Maintenance of Porous Asphalt and Its Function for Chloride Source Control." *ASCE Journal of Transportation Engineering*, American Society of Civil Engineers, Reston, VA.

Smith, D. R., Earley, K., and Lia, J. M. (2012). "Potential Application of ASTM C1701 for Evaluating Surface Infiltration of Permeable Interlocking Concrete Pavements." *Pervious Concrete Symposium: ASTM International*, accepted for publication in the *Journal of ASTM International*, Dec. 4, 2011, Tampa, FL.

Vancura, M., Khazanovich, L., and MacDonald, K. (2010). "Performance Evaluation of In-Service Pervious Concrete Pavements in Cold Weather—Final Report." Department of Civil Engineering, University of Minnesota, Minneapolis, MN.

Vancura, M., Khazanovich, L., and MacDonald, K. (2012). "Location and Depth of Pervious Concrete Clogging Material Before and After Void Maintenance with Common Municipal Utility Vehicles." *Proc., Transportation Research Board 91st Annual Meeting Compendium of Papers*, National Academy of Sciences, Washington, D.C.

Vanaskie, M. J., Myers, R. D., and Smullen, J. T. (2010). "Planning-Level Cost Estimates for Green Stormwater Infrastructure in Urban Watersheds." *Low Impact Development 2010: Redefining Water in the City 2010*, American Society of Civil Engineers (ASCE), Reston, VA.

9 Hydrologic and Hydraulic Design Methods

9.1 Introduction

Computational methods for hydrologic analysis, hydraulic functions, and treatment capacity of permeable pavements vary depending on the complexity of the design and level of detail in modeling the system properties. Chapter 1 presents permeable pavement systems designed to achieve a wide range of stormwater management design objectives including groundwater recharge, water quality, and volume and peak flow management. Practitioners use methods and models that range from generic infiltration systems to complex structures that address specific design objectives such as flow control or targeted pollutant removal. A number of ASCE publications cover the hydrologic response of innovative practices such as permeable pavement in detail (ASCE 1992).

This chapter does not direct the reader to use specific computational methods, but instead summarizes significant considerations required in computations for permeable pavement designs. The Low Impact Development (LID) Computational Methods Task Committee of the UWRRC/EWRI/ASCE, a sister task committee to the permeable pavements committee, continues to report on tools available for LID computations and provide application guidance.

The common variables used for quantifying permeable pavement hydrology are precipitation (rainfall), contributing runoff from adjacent areas (surface run-on), surface infiltration, surface runoff, evaporation, transpiration, storage volume, subgrade infiltration (exfiltration from the system), subdrain outflow, and

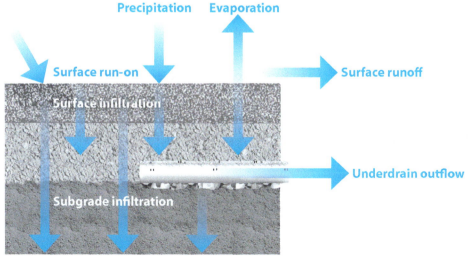

Figure 9-1
Water balance variables for permeable pavement
Source: © VHB

overflow (**Figure 9-1**). Key variables that control the hydrology and hydraulics of permeable pavements are defined in **Table 9-1**. As design specifications and objectives evolve, variables may become more complex in the characterization of permeable systems and may be included in expanded quantification methods. ertFor

Table 9-1 Key Variables Controlling Hydrology and Hydraulics of Permeable Pavement

VARIABLES	DEFINITION
Direct Precipitation	The amount of precipitation falling onto a permeable pavement surface
Surface Run-on	Runoff that flows onto a permeable pavement surface from adjacent areas
Surface Infiltration	Passage of direct precipitation and/or surface run-on through the permeable pavement surface layer
Surface Runoff	Water that flows off the permeable pavement surface and does not infiltrate into the base/subbase layers; runoff can occur if the rainfall intensity exceeds the surface infiltration rate, if the pavement base/subbase becomes saturated or if the pavement surface geometry allows for a direct runoff flow path
Evaporation/Transpiration	Conversion of water to water vapor that leaves the permeable pavement system through evaporation from the pavement material or through transpiration from surface vegetation that can be integral in some permeable pavement systems
Saturated Hydraulic Conductivity	A measure of the ability of water to move through a media under saturated conditions and pressure gradient
Hydraulic Head	Difference in elevation between two points in a fluid, producing pressure difference; the pressure influences both subgrade and infiltration subdrain outflow
Porosity	Void fraction (volume of voids/total volume)
Antecedent Moisture	Quantity of moisture present in the permeable pavement system before a rainfall event
Storage	Capacity of the system to store water in the void spaces in the pavement and base/subbase layers; may also include storage contained in underground vaults within the pavement subbase.
Subgrade Infiltration	Water that exits the subbase of a permeable pavement system and enters the soil subgrade
Subgrade Infiltration Rate	The infiltration rate of the soil underlying the permeable pavement system, which is a function of hydraulic head, saturated hydraulic conductivity, and soil moisture; the minimum rate of infiltration is the saturated hydraulic conductivity
Subdrain Outflow	Water exiting via underdrains
Overflow	Diversion of flow that exceeds the design volume of the system

example, the characterization of permeable pavement system outflow differs for full-, partial- or no-infiltration systems. The presence and configuration of any underdrains and overflows affect pavement storage volumes and discharge rates. Also, some designs include a filtering layer to specifically increase pollutant removal. Inclusion of this layer affects water quality and hydraulic performance. Depending on the methodology and modeling tools, the effects of such variables help determine the quantity and quality of the flow infiltrated to the natural soil, stored in the system, or discharged at a designed rate.

9.2   Hydrology and Hydraulics

Hydrology

The hydrologic processes mathematically characterized by most hydrologic models typically need some adjustments for modeling permeable pavement. In recent years, conventional methods have been adapted to simulate the minimal runoff, rapid infiltration, and temporary storage in a permeable media. In an ideal condition, permeable pavements mimic pervious surfaces by infiltrating flow and reducing runoff. With adequate base and subbase storage design, permeable pavements may perform accordingly. However, depending on storm intensity and recurrences, certain rainstorms may exceed the infiltration capacity of the system.

The pavement may then perform somewhere in between a pervious and impervious surface. Modeling this dual role can be a challenge. This creates some questions on how to model it as the permeable pavement is defined as a land cover, but it also functions as a part of the overall stormwater drainage/conveyance processes. Other common questions raised by the design community include how to assign runoff curve numbers (CNs), runoff coefficients, percent imperviousness, evapotranspiration rates, and time of concentration (TC) to reflect the effects of permeable pavements within the context of commonly used hydrologic modeling methods. Different regions, municipalities, or regulatory agencies may prescribe specific modeling methods. The following sections describe some of these parameters and their use in modeling approaches.

Curve Numbers

Curve numbers (CNs) characterize runoff from specific land uses (cover types) given the hydrologic soil group (HSG) and rainfall. Modeling methods using CNs are commonly used to estimate flow volumes and rates of runoff and are often the basis for the design of stormwater drainage/management systems. There are a couple of approaches to using CNs for permeable pavements:

Modified Curve Numbers

One interpretation of utilizing a CN based methodology is to use a modified CN to mimic the infiltration of a permeable pavement system. This approach is most applicable for permeable pavement systems in high infiltration soils. The pavement surface is characterized as a land cover type with a lower CN (resulting in less runoff) that mimics the process of the infiltration that occurs in the permeable pavement system. In this situation, the effects of the reduced volumes of runoff to the drainage system are modeled using the modified CN. The modified CN is a function of a specific design rainfall event . The user must determine the best representation given soil conditions and rainfall patterns. A 2-year design rainfall event will yield a different modified CN than a 100-year event.

McCuen (1983) presents the following **Equation 9-1** to determine the modified CN.

Equation 9-1: $CN^* = 200/[(P+2\Delta Q+2) - \sqrt{(5P\Delta Q+4\Delta Q^2)}]$

Where:
P = Design rainfall depth in inches
ΔQ = Post-development runoff depth minus the runoff depth stored by the infiltration practices in inches

This approach can enable the assessment of the reduction of runoff volumes that typically would have been directed to the closed pipe drainage system. This type of analysis is commonly completed in hydrologic/hydraulic modeling programs such as the Natural Resources Conservation Service (NRCS) TR-20 modeling program (Project Formulation Hydrology) by comparing pre-development hydrographs with the post-development hydrographs and comparing the volumes, peak flow rates, and timing of peak flows. The output hydrograph provides a visual and quantitative summary of the impact on volumes and peak flows as a result of the permeable pavement infiltration. Lower CNs are used to represent lower runoff that is a result of the pavement storage and infiltration. The results can help the designer evaluate the potential to reduce or eliminate closed pipe drainage systems with the use of permeable pavement systems.

Many models based on TR-20 methodology also typically have the capability of including infiltration systems, detention, retention facilities, and conveyances that characterize the infiltration function of a permeable pavement system. If these were used, the runoff would likely be directed to these systems/facilities in the model and the modified CN value would not be utilized.

The modified CN method is most appropriate for aggregating a large number of small permeable pavement areas, where the storage routing method would be impractical. Better results would be obtained if the source to treatment areas were similar for all nodes. For example, the method may be systematically applied for residential land uses where each lot may have a permeable pavement driveway or similar infiltration practice and can be modeled as a percentage of the drainage area. The level of peak discharge reduction is achieved by distributing the runoff storage volume over the entire watershed area and is reflected by adjusting the CN. This method can also be adapted for use on small individual lots, such as permeable parking lots on small sites if the details of the site and the pavement system are provided. Design examples illustrating the Modified CN methodology can be found in McCuen (1983) and MDE (2009).

Effective CN for Water Quality or Small Storms

An alternative approach is to assign the pavement surface effective CNs that vary based on specific storm events. The effective CNs can account for all of the flows leaving the permeable pavement system, so they may be used for permeable pavement systems utilizing underdrain and/or overflow configurations.

The effective CNs vary based on the size of the storm under investigation. This overall approach uses the Small Storm Hydrology work by Pitt (1994), which recognizes that the NRCS method is not accurate for small storms below 50 mm (2 in.) of rainfall. There are many instances where discharges need to be calculated from smaller storms in order to size water quality practices and account for cumulative annual rainfall and recharge processes. In this respect, models designed to calculate peak flow rates and volumes for larger storm events are not well suited and such models require modifications or development of new ones.

In addition to the use of effective CNs for small storm events, other approaches for use with larger storms or based on pavement system design characteristics have been implemented. Using monitored rainfall-runoff relationships, Bean (2007) computed permeable pavement CNs for specific storm events and estimated a single pavement effective CN using storms greater than 50 mm (2 in.). Leming (2007) modeled (without monitoring) permeable pavement CNs and computed a range of modified CN for individual simulated storm events. In Leming's approach, one permeable pavement system was assigned different CNs that varied depending on the size of the precipitation event. Schwartz (2010) provides a methodology on determining effective CNs dependent on contributing drainage areas and pavement reservoir depth.

Several states, including Maryland and Rhode Island, use an effective CN approach as a tool for computing peak discharges from small storms, ranging from the water quality storm (typically 13 to 25 mm [0.5 to 1 in.]) up to a 1-year, 24-hour storm. The CN varies based on the thickness of the pavement reservoir and the hydrologic soil group (HSG) of the subgrade. For specific examples, the reader is directed to these state stormwater guidance manuals. See: http://www.mde.state.md.us/programs/Water/ StormwaterManagementProgram/MarylandStormwaterDesignManual/Pages/programs/waterprograms/ sedimentandstormwater/stormwater_design/index.aspx and http://www.dem.ri.gov/programs/benviron/ water/permits/ripdes/stwater/t4guide/desman.htm

Runoff Coefficients

The runoff coefficient represents the fraction of total rainfall that ultimately appears as runoff, after storage and interception. This coefficient is used to estimate runoff volumes and peak flow rates in the Rational Method. Permeable pavements will intercept and store rainfall, at least temporarily. The amount of storage depends largely on the base/subbase storage capacity, subgrade infiltration rate and whether or not an underdrain and overdrain are utilized. Since runoff leaving subsurface drains usually reconnects to a downstream surface drainage network, this volume is included in the total runoff quantification.

Errors of over and under prediction are easily introduced by using the runoff coefficient approach, particularly in cases of surface run-on. Many hydrology references list runoff coefficients for different types of surfaces and values are often accepted for design carte blanche without careful consideration to design factors that affect storage and runoff. This pattern can be transferred to permeable pavements and designers are cautioned against using this simplistic approach.

Percent Imperviousness

In some jurisdictions, regulatory requirements may restrict the amount of impervious surface on a site or in a watershed. In these situations, permeable pavements can be considered as a percentage pervious and a percentage impervious. The percentage should be based on monitored runoff relationships and will vary depending on the outflow configuration (full versus partial versus no infiltration design), base/subbase depth, and the subgrade infiltration rate. There may be some instances where the pavement can be considered 100% pervious. North Carolina Department of Environment and Natural Resources (NCDENR) assigns pervious/impervious percentages depending on whether or not the pavement is installed over HSG A/B soils or C/D soils and provided the system meets specific design conditions (NCDENR 2012).

Evaporation/Evapotranspiration

The evaporation/evapotranspiration process is particularly relevant for permeable pavements in arid climates and for vegetated grid pavements. Computational methods that incorporate evapotranspiration rates, daylight hours, solar radiation, or other climate-dependent variables mimic the return of moisture to the atmosphere from soil and vegetation. Normally, more complex hydrologic modeling tools such as continuous simulation models can quantify the significance of evapotranspiration. Most models seldom consider this variable because it is not fully understood within permeable pavements, nor assumed to be a significant hydrologic component. Additionally, studies have shown that evaporation rates can be significant when the base/subbase ponding level is close to the pavement surface. However, the evaporation rate rapidly decreases as the water level drops (Li 2012; Nemirovsky 2013).

A study in North Carolina suggested that grid pavers filled with sand were able to remove up to 30% of the mean rainfall depth through water retention in the sand and subsequent evaporation during dry periods (Collins 2008). A study by Li (2012) in Davis, California indicated that micro-climatic thermal reductions are possible if the water level is kept near the surface of the permeable pavement through infusions of waste water such as waste landscape irrigation. Li indicates that lowered near-surface temperatures are greater outside of semi-arid climates where evaporation is not as rapid. Future research may prove that evaporation could be a significant hydrologic variable in specific regions and result in modifications to current modeling approaches.

Time of Concentration

When runoff enters a permeable pavement system that does not completely infiltrate into the soil subgrade, the water travels through the different permeable pavement layers before filling up the voids of the stone reservoir and discharging via an underdrain or overflowing. On site scale applications, this can result in an increase in runoff travel time and time of concentration for flows that ultimately are discharged from the system. In some instances, it may be necessary to use an extended time of concentration to reflect the peak delay resulting from permeable pavement runoff retention. The extended travel time through a permeable pavement system can be determined by storage routing and development of an outflow hydrograph, which is discussed below.

Hydraulics

Most permeable pavement system designs involve hydraulic analysis due to their functionality as pavements with subsurface storage and discharge capacity. Defining the routing process, controlling the discharge rate, specifying an outlet structure, and identifying similar parameters are essential in system hydraulics modeling. There are various computational techniques for routing flows through channels or storage systems with components that are applicable to the processes in permeable pavement systems. Rate of discharge over time, dewatering of the system, and addressing overflow mechanisms are also issues generally addressed by conventional hydraulic computational methods. These are discussed below in the context of modeling permeable pavements.

Storage Capacity of the Base/Subbase Materials

Storage is a dynamic process in permeable pavement systems; however, in some cases it is helpful to consider the static storage properties of the base/subbase materials. Doing so allows for a conservative assessment of storage capacity during rapid rainfall events or during minimal or blocked outflow (worst case) conditions.

The storage capacity of a static permeable pavement system includes the void spaces in the pavement, base/subbase layers, and storage contained in underground vaults within the pavement base/subbase. Assuming static conditions (no subgrade infiltration or outflow) and a level subgrade, the effective depth of storage can be computed using **Equation 9-2**. For example, if a reservoir has an effective porosity of 0.4 (40% voids), then a 25 cm (10 in.) thick base/subbase can hold 10 cm (4 in.) of runoff.

Equation 9-2: $d_r = d_p * \eta_r$

Where:
d_r = Depth of runoff stored
η_r = effective porosity of the reservoir layer
d_p = Depth of the reservoir layer

As discussed in **Chapter 1**, permeable pavements designed on a slope will have a lower storage capacity than a flat permeable pavement with the same reservoir design (**Figure 9-2**). The amount of storage reduction will be dependent on the surface and subgrade slopes and the specific configuration of the subbase (geometry, use of baffles, etc.).

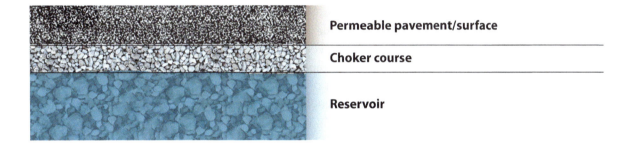

Permeable pavement/surface

Choker course

Reservoir

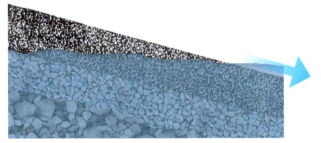

Figure 9-2
Difference in storage capacity for permeable pavements constructed on a flat surface versus a sloped surface.
Source: © VHB

Storage Routing

The net storage capacity of a pavement system is dynamic. Storage in permeable pavement is a function of the runoff inflow rate (from rainfall or surface run-on), the available void space, depth of the base/subbase materials, any runoff accumulated in the base/subbase from previous rainfall, the subgrade infiltration rate, and discharge through underdrain pipes and overflows. The storage follows a basic water balance equation (**Equation 9-3**):

Equation 9-3: $\Delta Storage = inflow + outflow$

Computational methods and routing techniques, such as dynamic storage-indication routing, are used to address the storage and hydraulic processes in the system. These methods for the most part can be applied to permeable pavement design. One approach is to model the permeable pavement system as a detention or recharge system with modest infiltration and/or controlled outlet(s) for discharge. The major point of difference is the presence of material in the storage reservoirs and the various layers of permeable pavement designs. Some modeling tools have incorporated open storage chambers within an aggregate bed (HydroCAD 2011) for additional water storage.

Consideration should be given to the following discharge and infiltration parameters in storage routing methods:

Subgrade Infiltration Rate

In a full or partial infiltration design, the subgrade infiltration rate of the permeable pavement system will need to account for the infiltration rate of the underlying soils. As discussed in **Chapter 1**, a conservative infiltration reduction factor of 0.5 (safety factor of 2) should be applied to the field measured infiltration rate to account for compaction that may occur during construction and sedimentation during the life of the pavement.

In a simplified approach, permeable pavement can be modeled as an infiltration basin. In the most basic infiltration models, runoff enters the pavement reservoir storage layer and a constant rate infiltration into the underlying subgrade is assumed. Physics-based models based on Green and Ampt or the Richards Equation can provide more accurate representations of varying infiltration rates (Lee 2011), but require more detailed information about the reservoir and underlying soil properties.

Outflow Through Perforated Underdrains

The physical configuration/elevations of underdrains affect the outflow rates and storage performance in permeable pavement systems. The three most common underdrain/outflow configurations designs with systems using perforated underdrain discharges include:

- **First case**—Perforated pipe placed at elevation above frequent water level in reservoir storage. The water level in the storage area during certain storm events is below the perforated pipe elevation with no discharge. This might be encountered in certain storm events if the reservoir is designed as an infiltration system and permeability of underlying soils are rapid, or the pipe is placed at a higher elevation to encourage a greater depth or ponding/greater head below the pipe to encourage higher rates of infiltration to the existing soils.

- **Second case**—Water is ponded above the perforated pipe while discharging. Flow is determined by the head of water above the pipe and the size(s) of the pipe(s) and its performance. In this condition the discharge rate can be manipulated by applying well-documented principles of orifice and pipe hydraulics to the perforated discharge pipe. This is the typical design scenario for an underground detention system (Ferguson 2005). It should be noted that, in most cases, the outflow rate from the underdrain will exceed system inflows. If desired, the underdrain can be fitted with a small orifice to control extended detention rates.

- **Third case**—The perforated pipe's capacity is large enough to allow free flow without limiting the discharge out of the reservoir. Instead, discharge is limited by the lateral flow rate through the reservoir aggregate; the pipe and its perforations carry water away as fast as the reservoir delivers it. In this condition the discharge rate is determined by the storage area's ponding depth, porosity, and hydraulic conductivity. It can be manipulated by controlling the number of pipes and the reservoir's hydraulics (Ferguson 2005).

Overflow Design

Sizing an overflow mechanism is a common design component of storage systems, addressing safe passage of flow exceeding the design capacity of the system. Bypass of larger flows in designing permeable pavements in particular is critical. The design should include the evaluation of the complete flow path during larger storm events, considering the flow path from point of overflow, including downstream receiving area and potential impacts.

Drawdown Time

Design dewatering times for permeable pavements typically range from 24 to 72 hours and may be a regulatory requirement or guideline. Designers should ensure that the total system outflow (overflow, pipe outflow, subgrade infiltration evaporation, and evapotranspiration) rate is such that the design dewatering times are met. **Equation 9-4** below, describes this relationship. The total outflow rate into the soil subgrade can be expressed as an average, variable or minimum rate, depending on the conservatism of the modeling approach.

Equation 9-4: Total system storage (VOL) = Total Outflow Rate (VOL/TIME)*Drawdown Time (TIME)

9.3 Water Quality

Water quality control in permeable pavements mostly depends on two pollutant removal pathways: infiltration and filtration. The effectiveness of these mechanisms depends on specific design features, e.g., depth of layers, media properties, native soils, and quality of inflow, among several other factors. Additionally, the use of vacuum sweepers and frequency of sweeping can further reduce pollutant loads. The primary function most commonly noted for water quality improvement in surface waters/resources is the reduction of total surface runoff volumes through infiltration. Water quality computational methods use the hydrology and hydraulic outputs for determining volume reduction. Many models can also consider pollutant build up, wash-off, transport, and decay mechanisms to quantify pollutant loading/load reductions. Selected mathematical methods can range from simple coefficients to complex chemical processes, including uptake, transport, and transition. Given the complexity of these variables, there is no single recommended model, practice, or data set for water quality modeling currently practiced. Research-based consensus on such tools for wider use is needed.

9.4 Modeling Tools

Stormwater modeling tools use static and dynamic characterizations of hydrologic and hydraulic processes to simulate rainfall runoff, discharge rate, storage, pollutant loading, and other processes. Depending on the model's algorithms, software programs vary from simple runoff calculations to dynamic water quantity and quality simulations over time. Each model serves a purpose depending on the scale, type, hydrology, or hydraulic processes being modeled. Their characteristics are grouped into general modeling categories summarized in **Table 9-2**.

Design storm models are rainfall event-based computational methods that predict the hydrologic response resulting from a storm event of a certain frequency. The primary variables in this type of modeling approach include; characterization of the storm event (typically volume and time distribution), and pre-defined soil and pavement responses to the rainfall event characterized by the use of runoff curve numbers. These models typically rely on readily available input data for the rainfall event characterization

Table 9-2 Examples of Different Types of Permeable Pavement Modeling Tools

DESIGN STORM	CONTINUOUS SIMULATION	DERIVED DISTRIBUTION	SPREADSHEET (WATER QUALITY)
Win-TR55	SWMM (XPSWMM, PCSWMM, InfoSWMM)	WinSLAMM	Simple Method
Win-TR20	HSPF	IDEAL	WTM
HydroCAD	SUSTAIN	CN Method	STEPL
Haestad Method (pond pack)	P8		LID Quicksheet
PerviousPave	BMPDSS		None
Permeable Design Pro	DRIP	DRIP	None

and may include or refer to standardized tables for defining cover conditions and soils conditions observed in the field, relating them to curve numbers or runoff coefficients. These approaches are often used for designing stormwater management systems for site development and for designing control structures for stormwater management practices. Two examples are the Rational Method and the NRCS TR-55 and TR-20 methods, which are integrated in a number of commercial software programs. These models can be used for the calculation of runoff flows on/into permeable pavements.

Continuous simulation models use series of historical precipitation records for estimating runoff volumes, storm durations, and pollutant loads. This method allows examination of watershed parameters and responses to long-term effects of storm events. These methods also allow for statistical evaluation of continuous and annualized rainfall/runoff events. Continuous simulation modeling that accounts for infiltration, evapotranspiration, depression storage, system storage, and detention can often provide the necessary tools to calculate hydrologic performance and characterize the processes that help achieve specific water quality objectives. This is especially true if the information is reviewed on an annual basis (i.e., annual volume reduction) rather than as an event-based performance analysis. These models can be used for watershed level analyses as well as assessment of individual BMPs within watersheds such as permeable pavements. For watershed level analyses, they can allow for detailed comparisons of alternative treatments to determine if watershed characteristics are maintained, improved and/or protected by those treatments (NYSDEC 2010).

Continuous simulation modeling allows for simulation of Low Impact Development (LID) techniques including permeable pavement and performance of flow duration analyses. For example, the Stormwater Management Model (SWMM 5.0.022) has an algorithm for data entry, process, and reporting related to LID practices including permeable pavements. Some of the LID controls in SWMM include LID storage layer (e.g., porosity, clogging factor, saturated hydraulic conductivity) or LID report variables (e.g., soil moisture, storage infiltration, evaporation) (US EPA 2011). An objective application of a continuous simulation modeling relies on a model validation process that results in a calibrated model for a specific watershed and incorporation of regional goals (James 2003; NYSDEC 2010).

Derived distribution models use relationships that predict the value of a dependent variable in terms of other independent variables. Such models combine hydrologic and hydraulic algorithms mentioned above. They rely on synthetic time series events, generally use mass balance for prediction of pollutant loading and are driven by empirical data and field observations. Such models offer calculation of runoff volumes and pollutant mass balances for different types of land use and land covers. For example WinSLAMM is designed to provide relatively simple outputs such as pollutant mass discharges and control measure effects for a large variety of potential conditions. (WinSLAMM 2011).

Spreadsheet models are simple models that evaluate individual, small-scale management practices, and sometimes evaluate the effects of them in a treatment train design. Such models are not designed to route flows dynamically, but are primarily water quality models intended to track pollutant sources and fates in urban and urbanizing watersheds. Spreadsheet models use simple coefficients and algorithms to calculate nutrient and sediment loads from different land uses and estimate load reduction based on management practice implementation. Several spreadsheet models such as the Watershed Treatment Model and StepL incorporate different types of permeable pavements for runoff reduction and treatment evaluation.

Permeable pavement specific models provide hydrologic and structural analysis for project-scale designs for (user designated) design storms. These are convenient tools since many local agencies require only site scale calculation for permeable pavement parking lots, alleys, or low speed roads. Industry-specific continuous simulation software models include PerviousPave issued by the American Concrete Pavement Association and Permeable Design Pro published by the Interlocking Concrete Pavement Institute. PerviousPave's hydrological design uses a modified version of the Los Angeles County method (LA 2002). The software determines the required minimum pervious concrete pavement thickness based on the design traffic, design life, and other structural inputs. The program also designs the required base/subbase reservoir thickness necessary to satisfy stormwater management requirements based on volume of water to be processed by the pavement within the required maximum detention time.

Permeable Design Pro provides structural design (base/subbase thickness) solutions using AASHTO 1993 flexible pavement design methodology that addresses traffic, design life, and layer coefficients. Based on user specified design storms, the program uses a dynamic infiltration, and detention model developed by the US Federal Highway Administration, Drainage Requirements in Pavement (FHWA 2002). This model characterizes the design of perforated pipes within the open-graded aggregate subbase, which facilitates applications in low-infiltration soils. Another software program that allows structural design for PICP in conjunction to the hydraulic design is LOCKPAVE Pro. (James 2003; Shackel 2006).

9.5   Agency Modeling Requirements

Design of stormwater practices, including permeable pavement systems, to meet the requirements of regulatory agencies often drives model selection. Some environmental or stormwater agencies have defined qualitative and quantitative criteria, and direct applicants to use common models. Other agencies require using agency designed models or specific regional models that incorporate acceptable variables and criteria to which the design must comply. While many of the variables in these models are covered in this chapter, agency-specific models, and regional requirements are outside the scope of this chapter.

9.6   Modeling Resources

BMPDSS (Prince George's County, MD's Best Management Practice – Decision Support System)
http://www.epa.gov/region1/npdes/stormwater/assets/pdfs/BMP-Performance-Analysis-Report.pdf

Haestad Method (PondPack) (Bentley Systems)
http://www.bentley.com/en-US/Products/Water+and+Wastewater+Network+Analysis+and+Design/stormwater.htm

HSPF (EPA's Hydrological Simulation Program)
http://water.usgs.gov/software/HSPF/

HydroCAD (Computer Aided Design tool for modeling stormwater runoff)
http://www.hydrocad.net/

IDEAL (StormOps)
http://www.stormopssoftware.com/

LOCKPAVE PRO, UNI-GROUP USA
www.uni-groupusa.org

LID Quicksheet (Milwaukee, WI MSD)
http://www.indy.gov/eGov/City/DPW/SustainIndy/WaterLand/Documents/Appendix%203%20MMSD%20Quicksheet.pdf

P8 (USEPA, Minnesota PCA & Wisconsin DNR's Program for Predicting Polluting Particle Passage thru Pits, Puddles, & Ponds)
http://wwwalker.net/p8/

Permeable Design Pro (ICPI)
http://www.icpi.org/node/1298

PerviousPave (ACPA)
http://acpa.org/PerviousPave/

PCSMWW (Computational Hydraulics International)
www.chiwater.com

STEPL (EPA's Spreadsheet Tool for Estimating Pollutant Load)
http://it.tetratech-ffx.com/steplweb/models$docs.htm

SUSTAIN (EPA's System for Urban Stormwater Treatment and Analysis Integration)
http://www.epa.gov/nrmrl/wswrd/wq/models/sustain/

SWMM (EPA's Storm Water Management Model)
http://www.epa.gov/athens/wwqtsc/html/swmm.html

WinSLAMM (Source Loading and Management Model)
http://winslamm.com/winslamm_updates.html

WTM (Center for Watershed Protection's Watershed Treatment Model)
http://www.cwp.org/documents/cat_view/83-watershed-treatment-model.html

Win-TR20 (USDA)
http://www.nrcs.usda.gov/wps/portal/nrcs/detailfull/null/?cid=stelprdb1042793

Win-TR55 (USDA)
http://www.nrcs.usda.gov/wps/portal/nrcs/detailfull/national/water/?&cid=stelprdb1042901

XPSWMM
www.xpsolutions.com

9.7 References

American Society of Civil Engineers (ASCE) (1992). "Design and Construction of Urban Stormwater Management Systems." *ASCE Manuals and Reports of Practice No.77*, WEF Manual of Practice FD-20, Urban Water Resources Research Council of ASCE and the Water Environment Federation, American Society of Civil Engineers (ASCE), Reston, VA, Water Environment Federation, Alexandria, VA.

Bean, E. Z., Hunt, W. F., and Bidelspach, D. A. (2007). "Field Survey of Permeable Pavement Surface Infiltration Rates." *Journal of Irrigation and Drainage Engineering*, American Society of Civil Engineers (ASCE), Reston, VA, 133(3), 249–255.

Collins, K. A., Hunt, W. F., and Hathaway, J. M. (2008). "Hydrologic Comparison of Four Types of Permeable Pavement and Standard Asphalt in Eastern North Carolina." *Journal of Hydrologic Engineering*. 13(12), 1146–1157.

Ferguson, B. K. (2005). Porous Pavements, CRC Press: Boca Raton, FL.

James, W., and von Langsdorff, H. (2003). "Computer-Aided Design of Permeable Concrete Block Pavement for Reducing Stressors and Contaminants in an Urban Environment." *Proc., 7th International Conference on Concrete Block Paving*, Sun City, South Africa, October 12–15, 2003.

Mallela, J., Larson, G., Wyatt, T., Hall, J. and Barker, W. (2002). "User's Guide for Drainage Requirements In Pavements." *Drip 2.0 Microcomputer Program*, Federal Highway Administration (FHWA),. FHWA-IF-02-05C, Office of Pavement Technology, United States Department of Transportation, Washington, DC.

Los Angeles County Department of Public Works (2002). *Development Planning for Storm Water Management*, Los Angeles County Department of Public Works, CA.

Lee, R. S. (2011). "Modeling Infiltration in a Stormwater Control Measure Using Modified Green and Ampt." *Master of Water Resources Engineering Thesis*, Department of Civil and Environmental Engineering, Villanova University, Villanova, PA.

Leming, M. L., Malcom, H. R., and Tennis, P. D. (2007). "Hydrologic Design of Pervious Concrete." Portland Cement Association and National Ready Mixed Concrete Association, Skokie, IL, Silver Spring, MD.

Li, H. (2012). "Evaluation of Cool Pavement Strategies for Heat Island Mitigation." *Ph. D. Dissertation*, Department of Civil and Environmental Engineering, University of California, Davis, CA.

McCuen, R. (1993). "Change in Runoff Curve Number Method." Maryland Department of the Environment, Baltimore, MD.

Natural Resource Conservation Service (NRCS) (2009). "Small Watershed Hydrology." *Technical Report 55 (WinTR-55)*, Natural Resource Conservation Service, USDA, Washington DC.

Nemirovsky, E. M., Welker, A. L, and Lee, R. (2013). "Quantifying Evaporation from Pervious Concrete Systems: Methodology and Hydrologic Perspective." *Journal of Irrigation and Drainage Engineering*. 139 (4) 271–277.

New York State Department of Environmental Conservation (NYSDEC) (2010). *New York State Stormwater Management Design Manual*, New York State Department of Environmental Conservation, Albany, NY.

New York State Department of Environmental Conservation (NYSDEC) (2008). *New York State MS4 General Permit (GP-0-08-002) Responsiveness Summary Appendix D*, New York State Department of Environmental Conservation, Albany, NY.

North Carolina Department of Environmental and Natural Resources (NCDENR) (2012). "Section 18 Permeable Pavements." *Stormwater BMP Manual*, North Carolina Department of Environmental and Natural Resources, chapter revised: Oct. 16, 2012, Raleigh, NC.

Pitt, R. (1994). "Small Storm Hydrology." University of Alabama—Birmingham, unpublished manuscript, presented at *Design of Stormwater Quality Practices*, Madison, WI, May 17–19, 1994.

Shackel, B. (2006). "Design of Permeable Paving Subject to Traffic." *Proc., 8th International Conference on Concrete Block Paving*, School of Civil and Environmental Engineering, University of New South Wales, San Francisco, CA.

St. John, M. S., and Horner, R. R. (1997). "Effect of Road Shoulder Treatments on Highway Runoff Quality and Quantity." Washington State Department of Transportation, WA-RD-429.1, <http://www.wsdot.wa.gov/Research/Reports/400/429.1.htm> (Apr. 29, 2008).

United States Environmental Protection Agency (US EPA) (1997). "Compendium of Tools for Watershed Assessment and TMDL Development." *United States Environmental Protection Agency (US EPA) Report*, EPA841-B-97-006.

United States Environmental Protection Agency (US EPA) (2011). "United States Environmental Protection Agency (US EPA) Stormwater Management Model (SWMM)." Version 5.0.022, Water Supply and Water Resources Division National Risk Management Research Laboratory, Cincinnati, OH.

10 Permeable Pavement Research Needs

10.1 Introduction

The implementation of permeable pavement installations and practices are expanding and becoming more commonplace. With this increased usage comes expanded and improved information on design, installation, and maintenance practices. While applied research and field experience have supported permeable pavement use and development over the past thirty years, there is a need for additional research, information, and data to support this trend and to aid the industry and users in realizing improved function, longevity, and verification design of goals. The following list of permeable pavement subjects are areas where additional research/information could benefit the implementation of the practice or further understanding of the effectiveness of the practice in achieving specific goals.

- Improved engineering specifications
- Validated hydrologic/hydraulic performance data modeling and techniques
- Validated and reliable structural design procedures
- Refined pollutant removal performance data and modeling techniques
- Initial and maintenance costs
- Installation and maintenance requirements for a range of applications
- Life-cycle costs and life cycle assessment
- Long-term evaluation studies
- Cold weather climate data and requirements
- Clogging/Hydraulic failure and rehabilitation

Each of these are discussed in the following sections.

10.2 Improved Engineering Specifications

There are no one-size-fits-all specifications and a significant amount of engineering judgment based on experience is required for the successful implementation of a project. There is a need for recommended engineering specifications updated as the practice improves.

Since local materials and permeable pavement uses vary across the country, construction specifications and design details prepared for each project that account for site specific and regional conditions and requirements are critical.

Technical institutes and industry associations representing permeable pavements such as the Interlocking Concrete Pavement Institute (Smith 2011a) and the American Concrete Institute (ACI 2010) have prepared valuable and complete technical specifications for using their specific systems. However, for many of the other pavement types, the specifications are more generic in nature. In addition, a gap commonly exists in where guidance and experience are required to transform generic recommendations into construction-ready specifications.

As described in **Chapter 7**, for pervious concrete and porous asphalt, engineering specifications that detail the specific product mixes and available product materials in the project area are critical. It is also critical that the pavement structural design is developed for the exact permeable pavement type and the specified use. Changes in materials or design requirements can change many factors of the pavement including strength, permeability, and the actual appearance of the final product.

This document provides some references for available specifications for engineers to review when preparing engineering documents for permeable pavements and also provides some outlines for the development of specifications that include recommended key items. The risk of problem projects and system failures could be reduced if the range of possible conditions were to be enumerated and proven decision paths were developed to aid in design and specifications. The goal is engineers would not be the very first or the very last to use design methods and specifications when developed and implemented. A process to develop some standard design specifications guidelines that can be tracked and modified as experience continues would be helpful.

10.3 Validated Hydrologic/Hydraulic Performance Data and Modeling Techniques

Computational Methods

There is an increasing need for engineers to have reliable and consistent methods for evaluating and modeling the hydrologic/hydraulic performance of permeable pavements. Accurate, validated system modeling is needed to assist in developing cost effective designs that can be expected to provide hydraulic functions as intended.

As noted in **Chapter 9**, engineers use several methods for modeling the hydrologic/hydraulic response that characterize permeable pavement performance. Methods vary among states, counties, and specific sites as do the rainfall patterns, geomorphology, soils, site specific features, and land use/cover. Therefore, engineers often rely on professional judgment (or agency requirements) to select methodologies and in some cases develop their own. In many cases, these models are theoretical characterizations without validation. The collection of hydrologic data from constructed sites to validate or adjust model configurations/inputs would contribute to better reliability of the results.

Further research is needed to predict permeable pavement hydrologic/hydraulic performance, based on materials, pavement type, surface and subgrade materials, maintenance, and infiltration rate reductions over a range of soil subgrade characteristics. Predicting pavement performance more accurately should increase its use and granting of stormwater mitigation credits. Curve number-based methods as well as dynamic time-based models call for further real-time validation and calibration.

Soil Hydraulic Capacity

Further research is required to evaluate the actual functioning hydraulic capacity of the underlying soil subgrade in permeable pavement systems.

Soil Subgrade Hydraulic Capacity

Until recently, permeable pavements were only recommended over coarse, rapidly draining soils to provide storage volume recovery and proper draining. This was evident in many state and local BMP manuals, which limited permeable pavement applications to soils with at least 13 mm/hr (0.5 in./hr) conductivity, and those limitations may have been borrowed from infiltration trench design guidelines. As the practice has expanded, there is more application of partial infiltrating systems in soils with low infiltration rates that incorporate detention/underdrains into designs. These demonstrate some water quality and runoff volume reduction benefits. Practices for enhancing infiltration from permeable pavements into underlying soils are being evaluated, including those with extended detention and increased head for recharge. Further study is required to assess their contributions.

Clogging of Soil Subgrades

Further research on long-term hydraulic behavior and clogging over time on various soil types is needed. Limited laboratory research by Mata (2009) indicates that sedimentation likely occurs at the top of the subgrade. The current practice is to use a safety factor of 0.5 applied to the measured soil infiltration rate to account for sedimentation as well as for construction-related compaction. Research needs to validate or revise this safety factor for various applications.

10.4 Validated, Reliable Structural Design Procedures

Engineers need standardized guidance on design procedures that result in reliable structural designs for permeable pavements.

Product Materials and Structural Performance

As noted in Chapter 7, it is critical for the success of permeable pavement projects to have engineering specifications that detail the specific mixes and product materials for the pavement type, region, and applied use. Each pavement has unique structural design requirements that can only be met when materials and mixes are specific to that pavement and address anticipated loads and intended structural design lives. The relationships among mix designs and structural performance require more research for permeable pavement.

Base/Subbase Materials and Structural Performance

Most permeable pavements use unstabilized open-graded aggregate materials for structural support. There is little data describing their structural performance, especially the open-graded ones like ASTM No. 2, 3, 4 or 57. Therefore, structural design methods are often theoretical and require validation using full-scale accelerated wheel load test methods similar to that accomplished for conventional pavements and bases. There have been theoretical studies by Caltrans (Jones et al. 2010) for porous asphalt and pervious concrete highway shoulders, which provided structural design charts for each. However, full-scale testing was the recommended outcome of these studies to validate the proposed design tables.

Open-graded materials tend to be more stress dependent, meaning that they stiffen with increased loads and such stiffness primarily depends on confining stresses, usually from adjacent aggregates, soils, curbs, and saturation. The shear behavior and stress-strain relationships under repeated loading need further research to characterize deformation rates. Larger sized open-graded aggregates tend to demonstrate greater stability than smaller size aggregates, hence their use in PICP and porous asphalt reservoir/subbase specifications.

While almost all permeable pavements use unstabilized open-graded aggregate bases/subbases, those stabilized with cement or asphalt have not been researched for effectiveness. Such materials are common in state DOT specifications and are typically used as drainage layers under conventional pavements. Structural testing of different permeable surfaces over stabilized bases may broaden permeable pavement applications to heavier load parking lot and road applications.

PICP Designs and Structural Performance

PICP also require the same type of full-scale structural performance testing to validate an AASHTO flexible pavement design-based design chart in industry literature (Smith 2011a). This chart assumes conservative values for soil subgrade strengths and for (unstabilized) bases with the maximum lifetime load of 1 million 80 kN (18,000 lb) ESALs. Base thickness design charts exist in Germany for up to approximately 800,000 10-tonne (22,000 lb) single axle loads (Zement 2003). United Kingdom PICP design charts use unstabilized and stabilized open-graded bases/subbases, and asphalt layers with drain holes to support traffic over 1.5 million 80 kN (18,000 lb) standard axles (Interpave 2008). Another approach has been taken in Australia and other countries in utilizing a fully mechanistic design method (Shackel 2006). While interlocking concrete pavements have been successfully used for port paving worldwide, the use of PICP in heavy-duty port and industrial pavement is a new application needing further testing and research (Sieglen 2004).

Pervious Concrete and Structural Performance

As mentioned in Chapter 3, the current structural design approach for pervious concrete assumes that methods used for designing rigid concrete pavement are transferable. Full-scale structural testing is needed to validate or modify this approach through mechanistic analysis. Since pervious concrete has compressive and flexural strengths similar to cement-treated bases, an alternative structural design method might consider a layer coefficient approach as used in flexible pavement design. Furthermore, pervious concrete requires development of test ASTM test methods to characterize strengths as inputs for structural design.

10.5 Refined Pollutant Removal Performance Data and Modeling Techniques

While the identified water quality benefits of permeable pavements support its increasing use, designers need methods that consistently and reliably predict pollutant removal capabilities or ranges depending on qualified conditions.

Aside from the surface sediment removal processes, additional water quality processes occur within the permeable pavement layer, base/subbase and the subgrade. The base/subbase materials act as a trickling filter with some benefits in reducing pollutants. While there are increasing number of studies on soluble pollutant removal rates and impacts to the groundwater, additional studies are needed, as described in the following sections.

Sediments/Particulate Matter/Attached Pollutants

Studies have shown that permeable pavements achieve the majority of their pollutant reductions by trapping sediments at the surface (Dierks et al. 2002). These sediments often have bound metals and pollutants attached to the sediment, which are also removed (Colandini et al. 1995; Kuang et al. 2007; Legret et al. 1999; Legret and Colandini 1999; Pagotto et al. 2000; Pratt et al. 1995; Rushton 2001). Additional information on expected removal rates for specific pavement use, maintenance and surrounding site conditions are needed.

Dissolved Metals

Dissolved metals are a class of the soluble pollutants affected by the subgrade, soil and filtering media associated with the permeable pavement design. A predominant factor for dissolved metal removal is the pH, specifically alkalinity. Some studies have shown that pervious concrete pavements can increase pH (Collins et al. 2008a) and alkalinity due to releasing $CaCO_3$, which can reduce metal mobility. Other studies have shown that concrete pavements buffer rainfall acidity. While some studies have been completed, further research could investigate the potential for greater dissolved metal removal by increasing contact time with cementitious materials and varying subsoils, possibly incorporating a limestone base/subbase or specific filtering courses into permeable pavement systems. Such research has implications on material selection, additives, system design, and using stratified layers of selected materials that enhance the overall pollutant removal capacity for metals and other pollutants. Some of these functions are discussed in Chapter 6.

Nitrogen and Phosphorus

Elevated nutrient concentrations such as nitrogen and phosphorus contribute to eutrophication and serve as the primary cause for many impaired waterways. Current data suggest that permeable pavement system performance varies widely for removal of nitrogen and phosphorus loads. With many TMDLs requiring nutrient reduction in stormwater runoff, permeable pavements provide a potential solution. However, additional data on the performance of nutrient removal is needed, especially from partial and no infiltration (lined) designs.

Nitrogen

Nitrogen is a concern in areas with shallow water tables or in areas with nutrient-sensitive groundwater or tidal wetlands/estuaries. This is especially true when nitrogen is in the form of nitrate which is soluble. Total nitrogen is comprised of NO_3-N and TKN, which in turn is comprised of NH_4-N and Org-N. Multiple concrete permeable pavement (PICP, pervious concrete and concrete grid) studies have found that NO_3-N in permeable pavement drainage was greater than asphalt runoff (Bean 2005; Bean et al. 2007a; Collins et al. 2008b; James and Shahin 1998). NH_4-N and TKN concentrations were lower than impervious asphalt runoff in multiple studies (Collins et al. 2008a; Bean et al. 2007a; James and Shahin 1998). Bean et al. (2007b) attributed these results to nitrification of NH_4-N into NO_3-N under aerobic conditions within the permeable pavement systems. However, other studies have reported that runoff from permeable pavements (Gilbert and Clausen 2006) and drainage from porous asphalt (Pagotto et al. 2000) had lower concentrations of all measured nitrogen species, including NO_3-N, than runoff from impervious asphalt pavements.

Future research should identify potential methods for reducing NO_3-N in discharges and the exfiltrate leaving permeable pavement systems. This may include developing anaerobic zones, especially within lined systems. Creating anaerobic zones within bioretention cells has been found to significantly reduce nitrogen loadings (Hunt et al. 2006). Creation of anaerobic zones within permeable pavement systems may be problematic, as the water chemistry could reduce certain pollutants, while increasing others, which may affect overall pavement performance as well. Research needs to determine the following:

1. The extent of biological activity within the pavement structure

2. Microfauna identification and means to enhance the nutrient removal capacity through system design

3. Maintenance to promote optimal biologic degradation of target pollutants while ensuring top performance and structural competence

4. The impact of the creation of anaerobic zone within the pavement layers

Phosphorus

Research results for phosphorus concentrations related to permeable pavement systems have been highly variable and are not well understood. Day et al. (1981) tested two concrete grid pavements and showed lower phosphorus concentrations than standard concrete runoff. Bean et al. (2007a) reported lower phosphorus species (orthophosphate, organic phosphorus and particle bound phosphorus) concentrations in permeable pavement leachate compared to asphalt runoff. While testing a third concrete grid type, Day et al. (1981), James and Shahin (1998), and Collins (2007) all reported greater concentrations of phosphorus in permeable pavement leachate compared to standard asphalt runoff. Improved understanding of the fundamental processes governing phosphorus dynamics would improve phosphorus predicted water quality impacts. Collins (2007) attributed increased phosphorus in leachate from permeable pavements to in-situ soils.

In these studies, it is not clear if phosphorus loads are originating from adjacent soils or other land uses. Research is needed to develop appropriate siting guidance to minimize phosphorus exports originating from adjacent soils and from contributing run-on from various land uses.

Filtering/Engineered Media

Media that chemically adsorbs dissolved pollutants may provide added water quality benefits, especially for non-infiltrating systems. Research into such materials will also have implications on the selection of materials, additives, system design, and the use of stratified layers of selected materials to enhance the overall pollutant removal capacity of the systems for metals and other pollutants. These media provide a charged surface that binds or precipitates dissolved charged pollutants. Incorporating these materials into pavement systems may improve pollutant retention (Kuang et al. 2007; Liu et al. 2005; Sansalone et al. 2007). There are many emerging studies showing promise for pollutant removal with engineered media including the removal of pathogens. However, there is limited information on the long term performance of the materials for continued pollutant removal. Additional research is needed in this area given the initial success in using engineered media to remove targeted pollutants and the limited long term performance data.

Thermal Effects

Increasing concern about pavement contributions to the urban heat island and thermal impacts to receiving waters has led to questions about possible permeable pavement mitigation of these effects. Pervious concrete has a higher specific heat and will maintain lower diurnal temperatures than other permeable or impervious pavements with higher material densities. This characteristic may be more important than surface reflectivity (albedo) in maintaining cooler surface temperatures than conventional concrete or asphalt. Regarding heated runoff, its reduction into surface water bodies, as well as the increase and increased cooler base stream flows (from infiltration) through open-graded bases/subbases, is one of these key benefits of permeable pavements. Additional research on methods for quantifying the benefits is needed. Additional research is also needed to develop design methods for permeable pavers and pavements to enhance reflectivity, especially within the context of reducing urban heat islands.

Another way to possibly mitigate thermal effects is by using PICP base as a thermal reservoir for ground source heat pumps. This approach needs additional research as horizontal ground source heat pumps are a well-established technology for building support.

Air Quality

Additional research may be warranted relative to photocatalytic coatings (typically TiO_2) on cement-based permeable pavement materials to reduce air pollution in urban settings.

10.6 Installation and Maintenance Requirements

There is a need for standardized guidelines for installation and maintenance requirements for a range of applications.

Certification Programs and Guidelines

While some permeable pavement materials suppliers and organizations have contractor certification programs on installation best practices, there is a need for some programs that include standardized guidelines regarding installation and maintenance practices. Knowledge of best practices improves with trial and error, and it would be helpful if the documentation of successful approaches is included in updated programs and guidelines as the practice expands and advances.

Pavement Maintenance and Management Practices as Applied to Permeable Pavements

Standard pavement maintenance and management is typically achieved through condition surveys and numerical indices that indicate when maintenance is required. Conventional asphalt and concrete pavements, as well as interlocking concrete pavements have such tools. There is ASTM D6433 *Standard Practice for Roads and Parking Lots Pavement Condition Index Surveys* and ASTM E2840 *Standard Practice for Pavement Condition Index Surveys for Interlocking Concrete Roads and Parking Lots*, which use the Pavement Condition Index originally developed by the US Army Corps of Engineers and promulgated by the American Public Works Association.

Research is needed that reviews the appropriateness of these standards to porous asphalt, pervious concrete and PICP. While many of the pavement distresses described in these standards may be similar, severities may be measured differently due to differences in surfacing materials as well as lower vehicle loads and speeds.

Infiltration Testing Standards for Permeable Pavements

In addition, there needs to be infiltration testing that guides when surface cleaning is needed. Originally pioneered by Bean (2005), ASTM C1701 *Standard Test Method for Infiltration Rate of in Place Pervious*

Concrete and ASTM C1781 *Standard Test Method for Surface Infiltration Rate of Permeable Unit Paving Systems* provide adequate and inexpensive test methods for pervious concrete. This test method has been used by the US EPA to demonstrate the applicability of this test method on porous asphalt and PICP (Borst et al. 2010; Smith et al. 2011b), and Bean (2005) also demonstrated its applicability to concrete grid pavements. ASTM C1781 *Standard Test Method for Surface Infiltration Rate of Permeable Unit Pavement Systems* has been recently developed to assess PICP and other segmental pavement surfaces. The results from this test method are comparable to ASTM C1701. Such standardization of surface infiltration test method or methods can provide equivalent maintenance performance comparisons. Recommended guidelines for the implementation of the tests and overall permeable pavement maintenance and management is needed.

10.7 Initial Costs, Life Cycle Costs and Life Cycle Assessment Data

There is a need for more information on the initial costs, maintenance costs and life cycle costs for permeable pavements, specifically as compared to standard pavements.

Cost Data Collection

There are hundreds of permeable pavements in use and there is a need to collect and coordinate cost information. With each project being unique, it is difficult to determine costs with limited data, and initial and maintenance cost documentation require systematic collection to be useful. When aggregated, such information would assist in developing project and system specific life-cycle costs. There are a sufficient number and age of all permeable pavement types to retrieve information from designers and project owners. This should be synthesized by pavement type, project size, and infiltration design, among other characteristics. Pavement condition can be surveyed to assist in estimating project life spans.

Life Cycle Assessment (LCA)

LCA, or the quantification of environmental impacts of manufacturing, construction, use, and decommissioning of permeable pavements is an emerging practice. Such impacts considered in a pavement LCA are found in *Methods, Impacts, and Opportunities in the Concrete Pavement Life Cycle* by Santero et al. (2011) and by Harvey et al. (2010). Sustainable rating systems such as LEED® v4, Greenroads, the Institute for Sustainable Infrastructure's Envision rating system and others encourage LCA. Most analyses quantify energy use and greenhouse gas emissions. Quantification of permeable pavement material and construction impacts is straightforward. The use phase, however, presents unique opportunities to find or develop tools that compare impervious systems to the reduced environmental impacts of permeable pavements. For example, tools are needed that measure lifetime water quantity and quality reductions and translate them into dollars saved and carbon reductions, not to mention social benefits. Such tools may consider land use impacts as well.

Green Infrastructure and Secondary Cost Benefits of Permeable Pavements

In addition to the typical intended functions, permeable pavements can also provide benefits that relate to costs, in the realm of sustainable urban design. These principles are articulated in the concept of "Complete Streets" (Smart Growth America, 2014). Such integrative approaches for the pavements that may account for social impacts from traffic calming, increased safety as a result of reduced accidents, increased property values and tax base, wayfinding design, and other benefits need research to demonstrate their fit and broader meaning to benefits to neighborhoods and cities. Dozens of cities have initiated green infrastructure programs. Permeable pavement designs are often integrated into green infrastructure/green street/green alley retrofits, achieving multiple benefits beyond reducing combined sewer overflows and water pollution. Such benefits are redefining the role of pavement in cities, but data is needed in order to properly quantify those benefits.

10.8 Long-term Evaluation Studies

There needs to be a coordinated collection of data on the long term performance of permeable pavements. Although permeable pavement systems at least 20 years old have been evaluated (Bean et al. 2007b) and have continued to function, monitoring studies generally focused on water quantity reductions have not exceeded eight years in duration. The US EPA permeable pavements study in Edison, NJ (Rowe et al. 2010) is expected to see monitoring for at least eight years and specifically focuses on monitoring stressors in soils. This type of evaluation of long-term impacts requires long-term provincial state or federal agency financial commitments to better understand performance over time. Long-term water quality impacts on specific pollutants and studies are also needed to better understand performance in relation to seasonal variations.

10.9 Cold Weather Climate Data and Requirements

Additional research to build on cold weather performance data such as deicer usage, impacts, and fate as well as pavement response to freeze/thaw cycles is needed. Data relative to the impact and cost of winter maintenance practices and equipment is needed.

Deicing

With the continued concern for limiting the use of deicers due to environmental concerns, pavement impact concerns, and costs, it is warranted to direct future research to consider quantifying deicing materials usage and their fate. Although Houle (2008) found that permeable pavement systems reduced deicing use by as much as 75%, impacts of infiltrated deicing materials on groundwater quality has not been well studied. Pavement color and sun orientation affect snowmelt rates and all permeable pavements reduce ice and slipping hazards. More studies on deicer use as well as its fate after movement through permeable pavement systems are needed.

Impacts of Deicers to Pavement System

Research should investigate the impacts deicing chemicals have on system performance and conditions these systems should not be used. For instance, there are questions about the impacts of deicers on pervious concrete and the methods for prevention of permeable pavement degradation. Results from these studies would have significant implications for cold climate installations. This is particularly the case with pervious concrete in that it has been subject to degradation and raveling from deicing materials. ASTM test methods for pervious concrete that help assure durability and performance in deicing should be a high priority.

Freeze/Thaw Impacts

Effects of freezing and thawing should be examined in future research to determine potential benefits or impacts on permeable pavement surface materials and functions. Information on the presence of pavement damage or lack of damage due to freeze/thaw conditions for permeable pavements as compared to standard asphalt pavements should be completed.

10.10 Clogging, Hydraulic Failure, and Rehabilitation Information

While strides have been made in the research relative to clogging and rehabilitation, this continues to be a major concern for those considering permeable pavement use. Additional research should be coordinated and disseminated to support its use.

Without proper maintenance, as with all surface filters, permeable pavements will clog given enough time and particulate loadings. Without maintenance, flow paths will become increasingly obstructed with foreign materials, reducing infiltration rates and increasing runoff rates (Colandini et al. 1995). However, studies have shown that even pavements considered clogged typically had infiltration rates equal or greater than rates assumed for HSG B soils 1.0 cm/hr (0.4 in/hr) (Bean et al. 2007b; Chopra and Wanielista 2007).

Material accumulation in void spaces reduces the size of drainage pathways in permeable pavements, restricting the flow rate and clogging the pavement (Bean et al. 2007a; Pratt et al. 1995; Kuang et al. 2007). Research has shown that initially, relatively large particles are trapped at the surface, while relatively small particles infiltrate into and possibly through the profile (Colandini et al. 1995; Mata 2009). However, as the drainage paths are obstructed, particles of decreasing diameter are trapped at the surface, creating a sediment/slime layer. The resulting decrease of infiltration rates has been shown to follow an exponential decay with time for given sediment load (Borgwardt 2006; Sansalone et al. 2007). Estimating sediment loading on soil surface, considering offsite runoff or not, is difficult and even more so for sediments migrating to the subgrade. Water and soil chemistry are also not fully understood on how sediments impact water quality performance and whether they should be considered as part of the design process.

As noted in the Chapter 8, removing the accumulation layer at the surface by vacuuming and/or pressure washing rejuvenates surface infiltration rates with varying degrees of success ranging from very good to poor (Bean et al. 2007b; Chopra and Wanielista 2007; James and Gerrits 2003; Kresin et al. 1996; Kuang et al. 2007). Future research should continue investigating the effects of rejuvenation methods.

Clogging is not limited to the surface. The storage layer and the soil subgrade interface sometimes include a geotextile. Fine sediments can migrate through the storage volume and accumulate at the surface of the geotextile, limiting infiltration, and recovery of the storage volume. Additional forensic research is needed to quantify the prevalence and identify contributing factors of geotextile clogging and design advantages or disadvantages associated with the use or absence of geotextiles.

10.11 Additional Topics

The following list of related topics were identified as areas where additional research/data collection would be of great benefit.

- Long term durability and performance of pavement surface and infiltration capacity
- Impact and cost of winter maintenance equipment
- Potential impacts of freeze/thaw conditions

- Evaluation of potential impact of icing conditions
- Use of permeable pavements for highway shoulder applications, including:
  - Effectiveness of surface infiltration under sheet flow conditions from adjacent roadway lanes
  - Potential impact of moisture conditions on adjacent pavement subgrade
  - Investigation of methods for managing water transfer to and from the stone reservoir below the shoulder and the traditional subbase
  - Stability of shoulders during emergency pull off conditions, particularly for heavy vehicles
  - Impact of potential traffic under temporary detour conditions
  - Recyclability of the shoulder materials
- Potential risk or benefits due to chemical spills
- Infiltration testing techniques for all surface types
- Evaluation of permeable asphalt mix designs and admixtures for heavier loadings and long term durability
- Development and adaptation of a field of QC standards for compaction testing of permeable asphalt
- Development and adaptation of a laboratory QC standard for pavement durability testing
- Evaluation of compaction methods for porous asphalt pavements and consolation of pervious concrete
- Development of vacuum sweepers and sweeping methods specifically for the high volume maintenance of permeable pavements associated with widespread implementation
- Recommended curing time for permeable asphalt and concrete as a function of pavement strength and temperature
- Study of road sanding and real versus perceived impacts on permeable pavements
- Development of simplifying assumptions to streamline the combined structural/hydrologic design process, where possible
- Correlation of site factors (e.g., landscaped area, trees, roadway AADT) and design factors (e.g., tributary area ratio, permeable pavement surface type) with the required frequency of vacuum sweeping
- Effect of chloride deicers on curing time and strength of pervious concrete; evaluation of alternative deicing materials
- Additional research on the binder draindown process and management for porous asphalt

10.12 Conclusion

The installation of permeable pavement systems can only be undertaken after in-depth site analyses. There is no single best approach. Each site must be evaluated individually. While the literature reflects many completed trials, field studies, and laboratory tests, there are still many research needs that could improve permeable pavement design, performance, and implementation. Fortunately, studies are underway that address many of the topics as previously discussed. The knowledge base will continue to grow as more research is published and as more data is collected on permeable pavement installations. With the results of new research in hand, permeable pavement technology will continue to improve and evolve.

10.13 References

American Concrete Institute (ACI) (2010). *Report on Pervious Concrete,* ACI-522R-10, American Concrete Institute, Farmington Hills, MI.

Bean, E. Z. (2005). "A Field Study to Evaluate Permeable Pavement Surface Infiltrations Rates, Runoff Quantity, Runoff Quality, and Infiltrate Quality." *MS thesis*, North Carolina State University, Department of Biological and Agricultural Engineering, Raleigh, NC.

Bean, E.Z., Hunt, W.F., and Bidelspach, D.A. (2007a). "Evaluation of Four Permeable Pavement Sites in Eastern North Carolina for Runoff Reduction and Water Quality Impacts." *Journal of Irrigation and Drainage Engineering*, 133(6). 583–592,133(6), 593– 601.

Bean, E. Z., Hunt, W.F., and Bidelspach, D.A. (2007b). "Field Survey of Permeable Pavement Surface Infiltration Rates." *Journal of Irrigation and Drainage Engineering*, 133(3), 247–255.

Borgwardt, S. (2006). "Long-term In-situ Infiltration Performance of Permeable Concrete Block Pavement," *Proc., 8th International Conference on Concrete Block Paving*, San Francisco, California, Interlocking Concrete Pavement Institute Foundation for Education and Research, Chantilly, VA.

Borst, M., Rowe, A. A., Stander, E. K. and O'Connor, T. P. (2010). "Surface Infiltration Rates of Permeable Surfaces: Six month Update (November 2009 through April 2010)." Water Supply and Water Resources Division, National Risk Management Research Laboratory, United States Environmental Protection Agency (US EPA), Edison, NJ.

Collins, K. A. (2007). "A Field Evaluation of Four Types of Permeable Pavement with Respect to Water Quality Improvement and Flood Control." *Thesis*. North Carolina State University, Raleigh, NC.

Collins, K. A., Hunt, W. F., and Hathaway, J. M. (2008a). "Hydrologic Comparison of Four types of Permeable Pavement and Standard Asphalt in Eastern North Carolina." *Journal of Hydrologic Engineering*, 13(12), 1146–1157.

Collins, K. A., Hunt, W. F., and Hathaway, J. M. (2008b). "Side-by-side Comparison of Nitrogen Species Removal for Four Types of Permeable Pavement and Standard Asphalt in Eastern North Carolina." *Journal of Hydrologic Engineering*, 15(6), 512–521.

Chopra, M., and Wanielista, M. (2007). "Report 2 of 4: Construction and Maintenance Assessment of Pervious Concrete Pavements" *Performance Assessment of Portland Cement Pervious Pavement*, Stormwater Management Academy, University of Central Florida, Orlando, FL, R. Browne, ed., final report FDOT Project BD521-02,<http:// www.dot.state.fl.us/research-center/completed_proj/summary_rd/fdot_bd521_02_rpt2.pdf>.

Colandini, V., Legret, M., Brosseaud, Y., and Balades, J. D. (1995). "Metallic Pollution in Clogging Materials of Urban Porous Pavements." *Water Science and Technology*. 32(1), 57–62.

Day, G. E., Smith, D. R., and Bowers J. (1981). "Runoff and Pollution Abatement Characteristics of Concrete Grid Pavements." *Bulletin 135*, Virginia Water Resources Research Center, Virginia Polytechnic Institute and State University, Blacksburg, VA.

Dierkes, C., Kuhlmann, L., Kandasamy, J., and Angelis, G. (2002). "Pollution Retention Capability and Maintenance of Permeable Pavements." *Proc., 9th International Conference on Urban Drainage, Global Solutions for Urban Drainage*, Portland, OR, American Society of Civil Engineers, Reston, VA.

Gilbert, J. K. and Clausen, J. C. (2006). "Stormwater Runoff Quality and Quantity from Asphalt, Paver, and Crushed Stone Driveways in Connecticut." *Water Research*, 40, 826–832.

Harvey, J., Kendall, A., Lee, I. S., Santero, N., Van Dam, T., and Wang T. (2010). "Pavement Life Cycle Assessment Workshop: Discussion Summary and Guidelines." UCPRC-TM-2010-03, University of California, Davis, CA.

Houle, K. M. (2008). "Winter Performance Assessment of Permeable Pavements a Comparative Study of Porous Asphalt, Pervious Concrete, and Conventional Asphalt in a Northern Climate." *Master's Thesis*, University of New Hampshire, Durham, NH.

Hunt, W. F., Jarrett, A. R., Smith, J. T., and Sharkey, L. J. (2006). "Evaluating Bioretention Hydrology and Nutrient Removal at Three Field Sites in North Carolina." *Journal of Irrigation and Drainage Engineering*, 132(6), 600–608.

Interpave (2008). "Permeable Pavements—Guide to the Design, Construction and Maintenance of Concrete Block Permeable Pavements." Interpave, 5th Ed., Interpave, Leicester, United Kingdom.

James, W., and Gerrits, C. (2003). "Maintenance of Infiltration in Modular Interlocking Concrete Pavers with External Drainage Cells." *Practical Modeling of Urban Water Systems*, Monograph 11, W. James, ed., Computational Hydraulics International, Guelph, Ontario, 417–435.

James, W., and Shahin, R., (1998). "Chapter 17: Pollutants Leached from Pavements by Acid Rain." *Advances in Modeling the Management of Stormwater Impacts*, W. James, ed., Guelph, Ontario, 6, 321–349.

Jones, D., Harvey, J., Li, H., Wang, T. Wu, R., and Campbell, B. (2010). "Laboratory Testing and Modeling for Structural Performance of Fully Permeable Pavements Under Heavy Traffic." Final Report, Report No. CTSW-RT-10-249.04, Division of Environmental Analysis, Storm Water Program, California Department of Transportation, Sacramento, CA.

Kresin, C., James, W. and Elrick, D.E. (1996). "Observations of Infiltration Through Clogged Porous Concrete Block Pavers." *Advances in Modeling the Management of Stormwater Impacts*, W. James, ed., Computational Hydraulics International, Guelph, Ontario, 5, 191–205.

Kuang, X., Kim, J., Gnecco, I., Raje, S., Garofalo, G., Sansalone, J. (2007). "Particle Separation and Hydrologic Control by Cementitious Permeable Pavement." *Journal of the Transportation Research Board*, No. 2025, Washington, DC, 111–117.

Legret, M. and Colandini, V. (1999). "Effects of a Porous Pavement with Reservoir Structure on Runoff Water: Water Quality and Fate of Metals." *Water Science and Technology*, 39(2), 111–117.

Legret, M., Nicollet, M., Miloda, P., Colandini, V., and Raimbault, G. (1999). "Simulation of Heavy Metal Pollution from Stormwater Infiltration Through a Porous Pavement with Reservoir Structure." *Water Science and Technology*, 39(2),119–125.

Liu, D., Sansalone, J., Cartledge, F. (2005). "Comparison of Sorptive Filter Media for Treatment of Metals in Runoff." *Journal of Environmental Engineering*, 131(8), 1178–1186

Mata, L. A. (2009). "Sedimentation of Pervious Concrete Pavement Systems." *Doctorate dissertation*, PCA R&D Serial No. SN3104, North Carolina State University, Raleigh, NC, Portland Cement Association, Skokie, IL.

Pagotto, C, Legret, M. and Le Cloirec, P. (2000). "Comparison of the Hydraulic Behaviour and the Quality of Highway Runoff Water According to the Type of Pavement." *Water Research*, 34(18), 4446–4454.

Pratt, C. J., Mantle, J. D. G., and Schofield, P. A. (1995). "UK Research into the Performance of Permeable Pavement, Reservoir Structures in Controlling Stormwater Discharge Quantity and Quality." *Water Science and Technology*. 32(1), 63–69.

Rowe, A. A., Borst, M., O'Conner T. P., and Stander, E. K. (2010). "Permeable Pavement Demonstration at the Edison Environmental Center." *Proc., Low Impact Development 2010: Redefining Water in the City*, American Society of Civil Engineers, Reston, VA.

Rushton, B. T. (2001). "Low-impact Parking Lot Design Reduces Runoff and Pollutant Loads." *Journal of Water Resources Planning and Management*, 127(3), 172–179.

Sansalone, J., X. Kuang, X., and Ma , J. (2007). "An In-situ Permeable Pavement and Media System for Hydrologic, Particulate and Phosphorus Management." *Proc., 2nd Biennial Stormwater Management Research Symposium*, University of Central Florida, Orlando, FL.

Santero, N., Loijos, A., Akbarian, M., and Ochsedorf, J. (2011). "Methods, Impacts, and Opportunities in the Concrete Pavement Life Cycle." *Concrete Sustainability Hub*, Massachusetts Institute of Technology (MIT), Cambridge, MA.

Sieglan, W. and von Langsdorff, H. (2004). "Interlocking Concrete Block Pavements at Howland Hook Marine Terminal." *Proc., 2004 ASCE Ports Conference*. Houston, Texas.

Shackel, B. (2006). "Design of Permeable Paving Subject to Traffic." *Proc., 8th International Conference on Concrete Block Paving*, School of Civil and Environmental Engineering, University of New South Wales, San Francisco, CA.

Smart Growth America (2014). "National Complete Streets Coalition." <http://www.smartgrowthamerica.org/complete-streets/>.

Smith, D.R. (2011a). *Permeable Interlocking Concrete Pavements*, 4th Ed., Interlocking Concrete Pavement Institute, Chantilly, VA.

Smith, D.R., Earley, K., and Lia, J.M. (2011b). "Potential Application of ASTM C1701 for Evaluating Surface Infiltration of Permeable Interlocking Concrete Pavements." *Proc., Symposium on Pervious Concrete,* ASTM International, Tampa, FL,

Zement (2003). Regenwasserversickerung durch Pflasterflachen, Bauerberatung Zement, Zeent-Merkblatt Strassenbau, Bundesverband der Deutsche Zementindustrie e.V., Köln, Germany, <www.BDZement.de>.

A Common Concerns Regarding Permeable Pavements Fact Sheet

Common Concerns Regarding Permeable Pavements Fact Sheet

CLOGGING

Concern: Permeable pavements clog and fail.

Response: Properly installed and maintained permeable pavements should not experience surface clogging that results in complete failure. Typically, even with some clogging, permeable pavements continue to infiltrate at acceptable rates due to the typically high initial infiltration rates at installation and permeable bases/subbases. Some clogging will occur over time from natural deposition and material tracked on to the pavement from vehicles, but pavement cleaning with a vacuum sweeper a minimum of two times (2x) per year should help prevent this type of clogging. More severe clogging is typically caused by a lack of cleaning and from one or more of the following: fine soils or organic materials being tracked onto the pavement, fines introduced with runoff from adjacent areas, or sand accumulation from winter sanding for safety. Recent studies (see references in **Chapter 8**) show that even when pavements experience clogging due to accumulation of fines in the pores of pervious concrete and porous asphalt, significant rehabilitation can be achieved with washing and vacuum sweeping (see **Chapter 8**).

Recommendations to prevent clogging include:

- Perform vacuum sweeping two times per year.
- Prohibit runoff from adjacent areas on to the pavement, especially with nearby unstable soils.
- Avoid placing pavement in areas of exposed or fine soils where run-on may occur.
- Prohibit winter sanding and include signs to clearly depict the prohibition.
- Avoid areas where excessive sand deposition may occur
- Avoid designs where organic matter accumulation may occur (i.e., under or directly adjacent to trees or at the toe of vegetated slopes).
- Prohibit traffic/equipment on pavements during construction and prohibit until all soils are permanently stabilized.
- Avoid landscape maintenance practices that may be deposit soil or organic matter on the pavement.

COSTS

Concern: Permeable pavements are cost-prohibitive.

Response: Permeable pavement system costs vary depending on the type of pavement surface as well as the base/subbase depth and design. Pavers and pervious concrete are typically more expensive than asphalt. Prices for plant-prepared materials vary based on the distance from the supplier to the project and whether the materials in the specifications are available locally. While permeable pavements may need additional depth of base or subbase materials to support anticipated vehicular loads, this subbase typically includes a reservoir course for stormwater storage. The detained and/or infiltrated stormwater can result in site development and management cost reductions if stormwater basins or trenches are not needed on the property. Cost savings related to the potential elimination of drainage infrastructure such as curbs, catch basins, and pipes may also be realized when utilizing permeable pavements instead of typical impervious

Source: © VHB

Common Concerns Regarding Permeable Pavements Fact Sheet continued

pavement installations. Estimated price comparisons for porous asphalt versus standard asphalt and other permeable pavement options are shown below in **Table A-1**. This table is also included in **Appendix B**.

Table A-1 Permeable Pavements Surface Cost Comparison

POROUS PAVEMENT TYPE	TYPICAL INSTALLED COST ($/SF)	TYPICAL COST RANGE ($/SF)
Porous Asphalt (5 cm [2 in.] surface course, 7.5 cm [3 in.] ATPB)	$6.00	$4.00 – $8.00
Pervious Concrete (15.25 cm [6 in.])	$10.00	$8.00 – $14.00
Interlocking Permeable Pavers & Rigid Open Cell Pavers (including 5 cm [2 in.] bedding layer)	(small hand installation) $13.00	$10.00 – $20.00
	(large mechanical installation) $6.50	$5.00 – $10.00
Open Celled/Grid Lattice Paving Systems	$7.00	$5.00 – $9.00
Proprietary Porous Pavement Products	Vary by manufacturer	

Note: Based on 17 actual bids with unit materials costs for permeable pavements (excluding open celled/grid lattice) from projects 2011–2013. General Estimates for installed permeable pavement surfaces with no sub-surface storage. Prices vary greatly with pavement depth, base/subbase and drainage variations.
*Estimate provided by National Ready Mix Concrete Association 2013
Source: CH2M Hill, 2013

MAINTENANCE

Concern: Maintenance costs are high.

Response: If proper and regular maintenance is completed (typically sweeping two times per year with a vacuum sweeper and protection pavement) to prevent severe clogging, the cost of maintaining pavements should not be higher than regular pavements (see **Chapter 8**). In response to the NPDES stormwater requirements, many municipalities are including vacuum sweeping and other operation/maintenance practices that benefit permeable pavement into their stormwater management. In some cases, they are requiring private entities to maintain BMPs such as permeable pavement with a specific maintenance regime that includes efficient cleaners such as vacuum sweepers. Contractors and municipalities are switching to these more up-to-date practices. Therefore, the costs for recommended maintenance practices for standard pavements and permeable pavements are not expected to be significantly different.

Common Concerns Regarding Permeable Pavements Fact Sheet continued

COLD CLIMATES

Concern: Permeable pavements do not function well in cold climates.

Response: Porous asphalt, pervious concrete, and permeable interlocking concrete pavements continue to be used successfully in cold climate areas. The porosity of the base/subbase aggregates, combined with the surface course porosity, allow for expansion of water if there are freeze/thaw cycles, rather than creating and/or expanding cracks as is often the case with impervious pavements.

Recent findings at the University of New Hampshire Stormwater Center show that the open-graded aggregates in the base/subbase of porous asphalt allows for the infiltration of snowmelt during sunny/warmer winter days. The results in a reduced need for salt applications to deal with the freeze/thaw cycles of thin layers of ice on the surface of the pavements that often form after snowmelt, plowing, or winter rain events. The black color of the asphalt promotes snow and ice melting under sunny conditions, and the porous surface allows infiltration of melt water into the base/subbase course and soil subgrade depending on the design (see **Chapter 2**). Similar findings exist with dark-colored permeable interlocking concrete pavements tested at the University (see **Chapter 2**).

Sanding should be prohibited as it increases clogging. However, as discussed in above, clogged pavements can be rehabilitated with surface cleaning and vacuuming, making complete failure unlikely.

Salt and other deicing materials can result in breakdown of the Portland cement in pervious concrete. Therefore, deicing materials should be avoided or used sparingly. Latex and/or fiber additions as well as higher sand content can increase resistance to deicing materials. Such deicing materials do not impact permeable interlocking concrete pavers because they are low absorption concrete. In addition, the units drain melting snow and ice in the permeable aggregate joints and bedding that have no cement.

DURABILITY

Concern: Permeable pavements are not structurally adequate for truck traffic or heavy loads.

Response: The lack of fines and the different mix properties of porous asphalt and pervious concrete result in pavement materials with lower structural capacity than standard asphalt and concrete pavement. The traffic load bearing capacity of the overall pavement section can be increased by increasing the depth of the base/subbase aggregates with proper compaction and installation. If suitable base/subbase depth is possible, and/or required for the stormwater design, this increased depth will increase the load bearing capacity of the permeable pavement so it may accommodate some use by trucks.

Permeable pavements should not be used for areas with heavy traffic, frequent truck or bus traffic, or repetitive turning movements by trucks or buses in the same location. Some of the current recommended uses for permeable pavements include, but are not limited to: light commercial parking lots and light use roadways; parking and roadways in residential subdivisions; shoulders on light and medium roadways; parking areas that do not experience significant truck traffic, such as office parks or commuter lots where the cars stay all day;

Common Concerns Regarding Permeable Pavements Fact Sheet continued

churches and institutions where parking may not be full each day of the week; overflow parking areas; and seasonal recreation parking areas. See the individual chapters for the recommended use for each type of permeable pavement.

SOIL CONSTRAINTS

Concern: Permeable pavements cannot be used over soils with low permeability or contaminated soils.

Response: Both permeable pavements and standard pavements require a base/subbase that allows for proper stability, sufficient load bearing capacity for the intended use, porous properties, and drainage to prevent the collection of water beneath the pavement. Permeable pavement bases/subbases can be placed over low permeability soils, but for most of these applications, an underdrain system is required. Some treatment occurs when the stormwater goes through the permeable pavement and base/subbase, even if it is not infiltrated into the native soils beneath. Infiltration occurs even in slow draining soils due to the pressure head that can be built if infiltrated stormwater is allowed to collect in the reservoir course, just as it would in a detention basin or subsurface infiltration system with a low flow outlet.

The use of a liner over the contaminated soils and an underdrain within the liner facilitates the use of permeable pavements over contaminated soils. Some treatment occurs when the stormwater goes through the permeable pavement and base/subbase, even if it is not infiltrated into the native soils beneath. The depth of the pavement section and base/subbase aggregates designed by the engineer are based on the pavement type and load bearing requirements for the intended traffic. Sites with contaminated soils and regulations in the specific jurisdiction must be addressed by the design engineer.

GROUNDWATER CONSTRAINTS

Concern: Permeable pavements cannot be used in high groundwater areas due to groundwater contamination risks.

Response: The depth to groundwater from the subbase of the permeable section may be a concern with a local jurisdiction. Infiltration of stormwater from the base/subbase of the permeable pavement section and base/subbase storage section requires a minimum separation from the seasonal high groundwater table. An underdrain can be used, thereby not relying on the native soils for infiltration. High groundwater situations require proper pavement and roadway construction regardless of permeable or impermeable.

Studies have shown that permeable pavements actually provide treatment in the upper layers of the pavement for many stormwater pollutants and this practice does not result in an increased threat to groundwater over any other type of infiltrating practice. The actual pavement itself provides treatment and adsorption of metals, hydrocarbons, oil, and grease with little movement of materials through the subgrade. Insoluble contaminants, such as salt, are not treated/removed within the pavement or base/subbase, but this is similar for any infiltration practice and the issue should be evaluated and managed based on risk. As noted previously in the **Cold Climates** concern section, significant reductions in salt use have been noted for

Common Concerns Regarding Permeable Pavements Fact Sheet continued

permeable pavements in the northeast due to the melting and infiltrating properties of porous asphalt and deicers should be avoided or used sparingly on pervious concrete.

SPILLS

Concern: A hazardous materials spill will contaminate the groundwater.

Response: A liquid hazardous material spill could potentially contaminate the groundwater anywhere it occurs with or without the presence of permeable pavement. However, the risk of large hazardous materials spills are much higher on highways and highway ramps, where permeable pavements are not recommended, compared to light traffic roadways and lighter use parking lots. In the event of a spill in any location, the soils need to be removed and any groundwater contamination requires remediation. Should a spill occur on permeable pavements, it is localized and not distributed widely into downstream areas through the stormwater system.

SLOPES

Concern: Permeable pavements cannot be used for sloped parking lots.

Response: The maximum slope on the pavement surface is recommended to be 5%. However, steeper slopes can be accommodated with stepped storage designed beneath the pavement. Stepped designs with berms or other impermeable barriers in the base/subbase system can be implemented to facilitate infiltration for sloped systems. See **Chapter 1 and Chapter 9** for more information.

*This information is provided based on information reviewed up to January 2014. It is expected that these answers will be subject to change as more research and project implementation becomes available.

Source: © VHB

B Design and Performance Summary Tables

B.1 Overview

This appendix summarizes general information for each permeable pavement type in this document. The information is based entirely on the data available at the time of the publication and are consistent with the information presented in the individual pavement chapters. It is anticipated that some of the information contained in these tables will change as the technology advances.

B.2 Summary Tables

Table B-1 Pavement Applications

TYPES OF APPLICATIONS	POROUS ASPHALT	PERVIOUS CONCRETE	PICP	GRID PAVERS
Overflow Parking	Yes	Yes	Yes	Yes
Primary Parking	Yes	Yes	Yes	Yes w/ aggregates Limited w/ grass
Sidewalks and Pathways	Yes	Yes	Yes	Limited
Road/Highways	Limited	Limited	Limited	Limited
Access Drives/Ring Roads	No (for heavy traffic)	Yes	Yes	Yes
Loading Areas	No	Yes	Yes	No
Frequent Truck Traffic	Limited	Yes	Yes	Limited
Cold Climates	Yes, should avoid the use of abrasives on pavement surface	Yes, should avoid the use of deicers, especially during first year following installation	Yes, require only standard snow removal procedures	Yes, special snow removal procedures required for grass filled applications

Table B-2 General Pavement Properties

PROPERTIES	POROUS ASPHALT	PERVIOUS CONCRETE	PICP	GRID PAVERS
Surface	Black, open-graded mix	Gray, open-graded mix; can be colored	Various shapes and colors	Large, open void spaces can be filled with soil and grass or washed; high permeability aggregates
Surface Open Void Space	18%–25%	15%–25%	5%–15%	Concrete: 20%–75%
Initial Installation Permeability	430–1250 cm/hr (170-500 in./hr)	750–5,000 cm/hr (300–2,000 in./hr)	1,000–1,500 cm/hr (400–600 in./hr)	Sand fill: 75–100cm/hr (30–40 in./hr) Aggregate fill: 500–1,000 cm/hr (200–400 in./hr) Grass and sod fill: 3–5 cm/hr (1–2 in./hr)
Longevity	30 years	30 years	50 years	30 years
Maximum Contributing Drainage Area Ratio	1.5:1	2:1	5:1	1:1
AASHTO Layer Coefficient (Typ.)	0.4–0.42	None; follow NRMCA recommendations for minimum layer thickness	0.3	None; use coefficients of underlying base/ subbase

Table B-3 Installation and Material Specifications

SPECIFICATIONS	POROUS ASPHALT	PERVIOUS CONCRETE	PICP	GRID PAVERS
Installation Procedure	Hot or warm mix	Ready mix, volumetric mixer or precast	Hand or mechanical	Hand or mechanical
Curing Time	Minimum 48 hrs (7 days recommended)	Minimum 7 days	None	None; however, grass-filled pavers require time for grass to establish (6–8 weeks seeded, 2–3 weeks sodded)
Surface Layer	2.5–10 cm (1–4 in.) porous asphalt mix (thickness dependent upon traffic loads); underlain by optional asphalt-treated permeable base 7.5–15 cm (3–6 in.)	10–30 cm (4–12 in.) pervious concrete mix (thickness dependent upon traffic loads)	80 mm (3.15 in.) thick concrete pavers with ASSHTO No. 8, No. 9 or No. 89 stone for fill material; 60 mm thick pavers can be used for pedestrian applications	Concrete: minimum 80 mm (3.15 in.) thick; Plastic grids: Standard or recycled plastic materials
ASTM or Industry Standard for Surface Material	None	ACI 522 and 522.1	ASTM C936 or CSA A231.2	Concrete: ASTM C1319 Plastic: None
Bedding Course	None	None	Typically AASHTO No. 8 or No. 9	If dense graded base: 13 to 25 mm (0.5 to 1 in.) thick layer of bedding sand; If open-graded base: 25 mm (1 in.) thick layer of AASHTO No. 8 stone
Choker Course/Base	10–20 cm (4–8 in.) thick layer (often AASHTO No. 57)	None	Typically AASHTO No. 57, 10 cm (4 in.) thick	
Reservoir Course/ Subbase*	Typically AASHTO No. 2 or No. 3	Typically AASHTO No. 57	Typically AASHTO No. 2, No. 3 or No. 4	Typically AASHTO No. 2, No. 3 or No. 4
Filter Layer (Optional)	20–30 cm (8–12 in.) poorly graded sand between choker and reservoir layer; should be underlain by choker course layer of 8 cm (3 in.) thick pea gravel.	None		
Notes		Restrictions for cold weather placement of surface material		

*Thickness of reservoir course is dependent on hydrologic and structural design factors.

Table B-4 Estimated Construction Costs for Pervious Pavements

	PAVEMENT REPLACEMENT SQUARE FOOT COSTS 2007				
	Demolition and Excavation	Installation of Subbase	Pavement Costs	Square Foot Costs	Comments
Porous Asphalt	$3.39	$3.40	$2.01	$8.80	46–76 cm (18–30 in.) excavation/backfill 7.5 cm (3 in.) porous asphalt
Standard Asphalt	$2.13	$1.04	$1.32	$4.49	15 cm (6 in.) excavation/backfill 15 cm (6 in.) asphalt
Pervious Concrete	$3.64	$3.40	$7.10	$14.14	46–76 cm (18–30 in.) excavation/backfill 14 cm (5.5 in.) pervious concrete
Standard Concrete	$1.51	—	$3.42	$4.93	No new base material 15 cm (6 in.) reinforced concrete

	PAVEMENT REPLACEMENT SQUARE FOOT COSTS 2005				
	Demolition and Excavation	Installation of Base/Subbase	Pavement Costs	Square Foot Costs	Comments
Porous Asphalt	$2.75	$1.88	$1.87	$6.50	46 (18 in.) excavation/backfill 7.5 cm (3 in.) porous asphalt
Pervious Concrete	$3.19	$1.88	$6.34	$11.41	

Table B-4 shows the estimated costs for conventional and pervious pavements from projects at the San Diego County Operations Center in 2005, 2007, and as updated.
*Square foot cost are based on actual cost received by the County of San Diego.
Source: Based on data from Clingan 2008

Table B-5 Permeable Pavements Surface Cost Comparison

POROUS PAVEMENT TYPE	TYPICAL INSTALLED COST ($/SF)	TYPICAL COST RANGE ($/SF)
Porous Asphalt (5 cm [2 in.] surface course, 7.62 cm [3 in.] ATPB)	$6.00	$4.00 – $8.00
Pervious Concrete (6 in.)	$8.00*	$6.00 – $10.00
Interlocking Permeable Pavers & Rigid Open Cell Pavers (including 5 cm [2 in.] bedding layer)	(small hand installation) $13.00	$10.00 – $20.00
	(large mechanical installation) $6.50	$5.00 – $10.00
Open Cell/Grid Paving Systems	$7.00	$5.00 – $9.00
Proprietary Porous Pavement Products	Vary by manufacturer	

Note: Based on data provided by CH2M Hill for 17 actual bids with unit materials costs for permeable pavements (excluding open celled/grid lattice) from projects 2011–2013. General Estimates for installed permeable pavement surfaces with no sub-surface storage. Prices vary greatly with pavement depth, base/subbase and drainage variations.
*Estimate provided by National Ready Mix Concrete Association 2013

Table B-6 Water Quality Performance Summary

STUDY	TYPE OF PERMEABLE PAVEMENT	LOCATION	% REMOVAL EFFICIENCY (CONCENTRATION)*															
			NH$_3$	NO$_{3,2}$	TKN	TN	DIN	Dissolved P	Ortho COD	TPH TP	TSS	Cd (total)	Pb (total)	Zn (total)	Cu (total)	COD	TPH-D
Eck et al. 2012	PFC	Texas		-46.0	63.0					78.0	96.0		96.0	90.0	69.0	60.0	
Eck et al. 2012	PFC	Texas		-31.0	49.0			0.0		66.0	91.0		90.0	87.0	56.0	20.0	
Eck et al. 2012	PFC	Texas		-6.0	25.0					75.0	93.0		90.0	87.0	60.0	27.0	
Bean et al. 2007b	PICP	North Carolina	83.9	-46.7	60.2	42.1			42.1	63.4	33.3			88.1	61.5		
Brattebo and Booth 2003	Plastic grids with gravel	Washington												61.9	88.8		
Brattebo and Booth 2003	Plastic grids with grass	Washington												38.9	99.9		
Brattebo and Booth 2003	CGP with turf	Washington												64.4	83.3		
Brattebo and Booth 2003	PICP	Washington												68.5	89.2		
Collins et al. 2010	Pervious concrete	North Carolina	85.3	-151.7	42.1	-2.4											
Collins et al. 2010	PICP	North Carolina	85.3	-331	49.5	-39.5											
Collins et al. 2010	CGP with sand	North Carolina	88.2	-58.6	49.5	23.4											
Collins et al. 2010	PICP	North Carolina	85.3	-210.3	49.5	-11.3											
Dierkes et al. 2002	PICP with gravel	Laboratory										98.0	98.0	97.0	96		
Fassman and Blackbourn 2011	PICP	Auckland, New Zealand									49.0			85	58		

STUDY	TYPE OF PERMEABLE PAVEMENT	LOCATION	% REMOVAL EFFICIENCY (CONCENTRATION)*														
			NH₃	NO₃,₂	TKN	TN	DIN	Dissolved P	Ortho COD	TPH TP	TSS	Cd (total)	Pb (total)	Zn (total)	Cu (total)	COD	TPH-D
Gilbert and Clausen 2006	PICP	Connecticut	72.2	50.0	91.3					33.6	66.9		66.7	71.3	64.7		
James and Shahin 1998		Ontario, Canada															
Kuang et al. 2007																	
Lui et al. 2005		Laboratory															
Pagatto et al. 2000	Porous asphalt	Nantes, France	73.0	68.7	42.9						81.1	68.2	78.3	66.2	33.3		
Roseen et al. 2011	Porous asphalt	New Hampshire		-87.1						20.0	88.9			75.0			91.6
Roseen et al. 2009	Porous asphalt	New Hampshire					-34.7			24.8	96.4			80.0			99.3
Rushton 2001	Pervious concrete	Florida	0.0	48.0		28.4			55.3	56.0	31.5		57.1	35.9	57.7		
Rushton 2001	Pervious concrete	Florida	0.0	36.1		25.7			32.4	38.8	62.4		68.9	52.8	67.4		
Sansalone et al. 2007		Laboratory															
Toronto and Region Conservation Authority 2007	PICP	Ontario, Canada	-68.0		53.0				0.0	53.0	81.0			73.0	13.0	-188.0	
Yong et al. 2011	Porous asphalt	Laboratory				3.0				30.0							
Yong et al. 2011	PICP	Laboratory				-14.0				34.0							

STUDY	TYPE OF PERMEABLE PAVEMENT	LOCATION	% REMOVAL EFFICIENCY (CONCENTRATION)*														
			NH_3	$NO_{3,2}$	TKN	TN	DIN	Dissolved P	Ortho COD	TPH TP	TSS	Cd (total)	Pb (total)	Zn (total)	Cu (total)	COD	TPH-D
Number of Studies, n			9.0	14.0	11.0	9.0	1.0	1.0	4.0	12.0	12.0	2.0	8.0	17.0	15.0	4.0	2.0
MEDIAN			83.9	-46.3	49.5	3.0	-34.7	0.0	37.3	45.9	81.0	83.1	84.1	73.0	64.7	23.5	95.4
MEAN			63.7	-59.5	52.3	6.2	-34.7	0.0	32.4	47.7	72.5	83.1	80.7	71.9	66.5	-20.3	95.4
MAXIMUM			88.2	68.7	91.3	42.1	-34.7	0.0	55.3	78.0	96.4	98.0	98.0	97.0	99.9	60.0	99.3
MINIMUM			0.0	-331.0	25.0	-39.5	-34.7	0.0	0.0	20.0	31.5	68.2	57.7	35.9	33.3	-188.0	91.6

* Removal efficiencies listed are based on concentration (EMC)
**Efficiencies represent the median efficiency based on EMC
Note: Not all removal efficiencies from individual studies are statistically significant.

Table B-7 Hydrologic Performance Summary

STUDY	TYPE OF PERMEABLE PAVEMENT	LOCATION	% VOLUME REDUCTION*	NOTES
Anderson et al. 1999	PICP	Laboratory	30.0[1]	Pavement pre-saturated
Anderson et al. 1999	PICP	Laboratory	55.0[1]	Dry pavement
Bean et al. 2007b	PICP	North Carolina	66.0	
Bean et al. 2007b	PICP	North Carolina	100.0	Infiltration design, no underdrains
Collins et al. 2008	Pervious concrete	North Carolina	12.8	Located over clay soils
Collins et al. 2008	PICP	North Carolina	47.6	
Collins et al. 2008	CGP with sand	North Carolina	27.8	
Collins et al. 2008	PICP	North Carolina	3.1	
Dreelin et al. 2009	Plastic grids with grass	Georgia	93.0	Located over clay soils
Fassman and Blackbourn 2010	PICP	New Zealand	28.0	Located over impermeable soils
Gilbert and Clausen 2006	PICP	Connecticut	72.2	
Legret and Colandini 1999	Porous asphalt	France	96.7	
Roseen et al. 2011	Porous asphalt	New Hampshire	25.0[2]	Located over HSG C soils
Rushton 2001	Pervious concrete	Florida	29.1	
Rushton 2001	Pervious concrete	Florida	31.7	
MEDIAN			**39.7**	
MEAN			**50.7**	
MAXIMUM			**100.0**	
MINIMUM			**3.1**	

* Volume reduction values are based on comparison to standard asphalt, unless noted. Volume reduction compared to the volume leaving the permeable pavement system via underdrains and surface runoff.

[1] Compared to quantified influent

[2] Compared to calculated rainfall volume

B.3   References

Andersen, C. T, Foster, I. D. L., and Pratt, C. J. (1999). "Role of Urban Surfaces (Permeable Pavements) in Regulating Drainage and Evaporation: Development of a Laboratory Simulation Experiment." *Hydrological Processes*, 13(4), 597.

Bean, E. Z., Hunt, W. F., and Bidelspach, D. A. (2007). "Evaluation of Four Permeable Pavement Sites in Eastern North Carolina for Runoff Reduction and Water Quality Impacts." *Journal of Irrigation and Drainage Engineering*, 133(6), 583–592.

Brattebo, B. O., and Booth, D. B. (2003). "Long-term Stormwater Quantity and Quality Performance of Permeable Pavement Systems." *Water Research*, 37(18), 4369–4376.

Clingan. (2008). Personal email communication from Dane Clingan. San Diego County, CA.

Colandini, V., Legret, M., Brosseaud, Y., and Balades, J. D. (1995). "Metallic Pollution in Clogging Materials of Urban Porous Pavements." *Water Science and Technology*. 32(1), 57–62.

Collins, K. A., Hunt, W. F., and Hathaway, J. M. (2008). "Hydrologic Comparison of Four Types of Permeable Pavement and Standard Asphalt in Eastern North Carolina." *Journal of Hydrologic Engineering*.

Collins, K. A., Hunt, W. F., and Hathaway, J. M. (2010). "Side-by-side Comparison of Nitrogen Species Removal for Four Types of Permeable Pavement and Standard Asphalt in Eastern North Carolina." *Journal of Hyrdologic Engineering*. 15(6), 512–521.

Dierkes, C., Kuhlmann, L., Kandasamy, J., and Angelis, G. (2002). "Pollution Retention Capability and Maintenance of Permeable Pavements." *Proc., 9th International Conference on Urban Drainage*, Global Solutions for Urban Drainage. American Society of Civil Engineers (ASCE), Portland, OR.

Dreelin E. A., Fowler, L. Carroll, R. C. (2006). "A Test of Porous Pavement Effectiveness on Clay Soils During Natural Storm Events." *Water Research*, 40, 799–805.

Eck, B. J., Klenzendorf, J. B., Charbeneau, R. J., and Barrett, M. E. (2010). "Investigation of Stormwater Quality Improvements Utilizing Permeable Friction Course (PFC)." Texas Department of Transportation. Report No. FHWA/TX-11/0-5220-2/.

Eck, B. J., Winston, R. J., Hunt, W. F., and Barrett, M. E. (2012). "Water Quality of Drainage from Permeable Friction Course." *Journal of Environmental Engineering*. 138(2), 174–181.

Fassman, E. and Blackbourn, S. (2010). "Urban Runoff Mitigation By a Permeable Pavement System Over Impermeable Soils." *Journal of Hydrologic Engineering*. 15 (6), 475-485.

Fassman, E. A., and Blackbourn, S. D. (2011). "Road Runoff Water Quality Mitigation by Permeable Modular Concrete Pavers." *Journal of Irrigation and Drainage*, 137(11)

Gilbert, J. K. and Clausen, J. C. (2006). "Stormwater Runoff Quality and Quantity From Asphalt, Paver, and Crushed Stone Driveways in Connecticut." *Water Research*, 40, 826–832.

James, W. and Shahin, R. (1998). "Chapter 17: Pollutants Leached from Pavements by Acid Rain." *Advances in Modeling the Management of Stormwater Impacts*, W. James, ed., Guelph, Canada, 6, 321–349.

Kuang, X., Kim, J., Gnecco, I., Raje, S., Garofalo, G., Sansalone, J. (2007). "Particle Separation and Hydrologic Control by Cementitious Permeable Pavement." *Journal of the Transportation Research Board*, No. 2025, Washington, D.C. 111–117.

Legret, M., and Colandini, V. (1999). "Effects of a Porous Pavement with Reservoir Structure on Runoff Water:

Water Quality and Fate of Metals." *Water Science and Technology.* 39(2), 111–117.

Liu, D., Sansalone, J., Cartledge, F. (2005). "Comparison of Sorptive Filter Media for Treatment of Metals in Runoff." *Journal of Environmental Engineering*, 131(8), 1178–1186.

Newman, A. P., Pratt, C. J., Coupe, S. J., and Cresswell, N. (2002). "Oil Bio-degradation in Permeable Pavements by Microbial Communities." *Water Science and Technology*, 45(7), 51–56.

Pagotto, C, Legret, M. and Le Cloirec, P. (2000). "Comparison of the Hydraulic Behaviour and the Quality of Highway Runoff Water According to the Type of Pavement." *Water Research.* 34(18), 4446–4454.

Pratt, C. J., Mantle, J. D. G., and Schofield, P. A. (1995). "UK Research into the Performance of Permeable Pavement, Reservoir Structures in Controlling Stormwater Discharge Quantity and Quality." *Water Science and Technology*, 32(1), 63–69.

Pratt, C. J., Newman, A. P., and Bond, C. P. (1999). "Mineral Oil Bio-degradation Within a Permeable Pavement: Long Term Observations." *Water Science and Technology.* 39(2), 103–109.

Roseen, R. M., Ballestero, T. P., Houle, J. J., Briggs, J. F., and Houle, K. M. (2011). "Water Quality and Hydrologic Performance of a Porous Asphalt Pavement as a Stormwater Treatment Strategy in a Cold Climate." *Journal of Environmental Engineering*, In press.

Roseen, R. M., Ballestero, T. P., Houle, J. J., Avellaneda, P., Briggs, J. F., Fowler, G., and Wildey, R. (2009). "Seasonal Performance Variations for Stormwater Management Systems in Cold climate Conditions." *Journal of Environmental Engineering.* 135(3), 128–137.

Rushton, B. T. (2001). "Low-impact Parking Lot Design Reduces Runoff and Pollutant Loads." *Journal of Water Resources Planning and Management.* 127(3), 172–179.

Sansalone, J., X. Kuang, X., Ma , J., (2007). "An In-situ Permeable Pavement and Media System for Hydrologic, Particulate and Phosphorus Management." *Proc, 2nd Biennial Stormwater Management Research Symposium*, University of Central Florida, Orlando, FL.

Toronto and Region Conservation Authority (2007). "Performance Evaluation of Permeable Pavement and a Bioretention Swale." Seneca College, King City, Ontario. Interim Report #3.

Yong, C. F., Deletic, A., Fletcher, T. D., and Grace, M. R. (2011). "Hydraulic and Treatment Performance of Pervious Pavements Under Variable Drying and Wetting Regimes." *Water Science and Technology.* 64(8),1692–1699.

C Standards, Specifications, Testing Methods, Resources, and References

C.1 Pervious Concrete Standards, Specifications, and Testing Methods

Freshly Mixed Pervious Concrete Void Content

In 2008, the first standardized testing method was developed by ASTM International for acceptance of pervious concrete material from the mixing truck prior to placement, entitled ASTM C1688 *Standard Test Method for Density and Void Content of Freshly Mixed Pervious Concrete* (ASTM 2008). The ASTM website describes the test's significance and use as follows:

> *"This test method provides a procedure for determining the density and void content of freshly mixed pervious concrete. This test method is applicable to pervious concrete mixtures containing coarse aggregate with a nominal maximum size of 25 mm (1 in.) or smaller. The measured fresh density may be used as verification of mixture proportions. The fresh density and void content calculated from this test may differ from the in-place density and void content, and this test shall not be used to determine in-place yield."* (ASTM 2009a; Haselbach 2010)

Hardened Pervious Concrete Void Content

Pervious concrete quality and porosities are dependent on the mixture proportions in the truck and on field placement techniques. However, hardened pervious concrete exhibits different proprieties. Therefore, a test for hardened concrete was developed by ASTM. The ASTM C1754 *Standard Test Method for Density and Void Content of Hardened Pervious Concrete* applies to cored or molded cylinder test specimens taken from the project after hardening or curing.

Two drying methods labeled A and B are used for the hardened test specimens. Drying Method A uses a lower temperature to determine the constant dry mass of the pervious specimen. Depending on the initial condition of the pervious specimen, obtaining the constant dry mass may take as long as one week or more. Drying Method B uses a much higher temperature, and therefore the constant dry mass is attained much more quickly. Test specimens using Drying Method B are not used to determine other properties of the pervious concrete. If other physical properties such as strength or infiltration are determined from the specimens, Drying Method A should be used for testing. Drying Method B may produce lower densities and correspondingly higher void contents than Drying Method A. Results from the two should be treated separately and not combined.

Surface Durability and Strength Testing of Pervious Concrete

ASTM C1747 *Standard Test Method for Determining Potential Resistance to Degradation of Pervious Concrete by Impact and Abrasion* was developed to assess the raveling potential of pervious concrete mixes. Since raveling is a common failure mode, this test enables comparison of various mixtures resistant to raveling. The test method uses cylindrical specimens and abrades them in a specific equipment (used for Los Angeles abrasion loss) to determine the mass lost. The results may be used to compare proposed mixture proportions, but it is not intended for qualifying mixtures or jobsite acceptance testing. At this time, there is no correlation between this test method and field raveling resistance of pervious concrete. Field raveling is caused by improper paste consistency, workability loss, inadequate compaction, and improper curing. At this writing, ASTM is developing a compressive strength test method entitled *New Test Method for Compressive Strength of Pervious Concrete Cylinders made in the Laboratory.*

Field Infiltration Testing of Pervious Concrete

A major characteristic of pervious concrete is its ability to infiltrate water. In order to verify continued satisfactory infiltration performance, there is a need for standardized testing methods to measure the infiltration capability of pervious concrete in the laboratory and in the field (ACI 2010). A falling head permeameter system can be used to measure saturated hydraulic conductivity (KH) in laboratory samples (Montes and Haselbach 2006). The hydraulic conductivity test is an accepted, standardized method of quantifying permeability in granular soils. However, flow through pervious concrete rarely occurs under saturated conditions and the falling head test is somewhat complex and difficult to use for in-situ applications of pervious concrete due to leakage and other concerns. Saturated flow is commonly at a higher rate than unsaturated flow in porous media due to the effects of capillary pressure in unsaturated media (Haselbach 2010).

To address this situation, ASTM adopted a field infiltration rate testing method more representative of unsaturated conditions in the field entitled ASTM C1701 *Standard Test Method for Infiltration Rate of in Place Pervious Concrete* (ASTM 2009c). This test has a method for pre-wetting the pervious pavement. ASTM related studies (see **Section C.5 References**) have demonstrated that for unsaturated conditions and many pervious concrete samples, there are usually no differences between wet and dry conditions. However, there are some cases in which pre-wetting does make a difference. Wetted conditions are always tested. Dry and wet conditions may vary more in cases with low porosities or flow limited due to clogging, since the impacts of suction in the wetted smaller pores may have a larger relative contribution (Haselbach 2010). This testing method also provides alternatives for testing pavements with higher flows using more water than those with a lower range of infiltration rates. The former minimizes errors associated with timing; the latter minimizes the time and water volumes needed for testing. This test is recommended to determine if maintenance may be required due to a decrease in infiltration rates.

Specifications for Pervious Concrete

Additional references include the Pervious Concrete specification developed by the American Concrete Institute Technical Committee 522 (ACI 2008). This specification also provides a checklist for construction.

The Technical Committee on Transportation and Infrastructure of the ASCE Technical Council on Cold Regions Engineering has developed a publication entitled, *Monograph on Pavement Design in Cold Regions.* This includes guidelines for mix design and construction of pervious concrete in cold climates. See http://www.asce.org/Content.aspx?id=2147489074 for further information.

C.2 Porous Asphalt Standards, Specifications, and Testing Methods

Field Infiltration Testing of Porous Asphalt

The ASTM C1701 *Standard Test Method for Infiltration Rate of In Place Pervious Concrete* is also being applied to porous asphalt pavements.

Draindown Testing of Porous Asphalt

ASTM C6390 *Standard Test Method for Determination of Draindown Characteristics in Uncompacted Asphalt Mixtures* is used to determine the amount of draindown for a given uncompacted asphalt mixture at temperatures comparable to those during production, storage, transport, and placement. This test is primarily used for mixtures with high coarse aggregate content such as porous asphalt (Note: open-graded friction course is typically made of finer aggregates than ATPBs [asphalt treated permeable bases]) and stone matrix asphalt (SMA). An asphalt mix resistant to draindown helps prevent surface clogging. At this writing, ASTM is developing a construction guide for porous asphalt.

C.3 Permeable Pavers Standards, Specifications, and Testing Methods

Field Infiltration Testing of PICP

ASTM C1781 *Standard Test Method for Surface Infiltration Rate of Permeable Unit Pavement Systems* (ASTM 2013)was developed specifically for use on PICP, concrete grid pavements and other permeable segmental concrete pavement surfaces. This test method uses equipment and calculations identical to C1701 with testing procedures that guide users on locating the test equipment on a permeable paver pattern. Results from C1781 can be compared to test results from C1701.

C.4 Resources

Permeable Pavement Resource—General

Ferguson, B. K. (2005). *Porous Pavements,* CRC Press, Boca Raton, FL.

North Carolina Department of Environment and Natural Resources (NC DENR), "Subchapter 3.10. Design Installation, and Maintenance for Permeable Pavement Systems." Permeable Pavement Systems, Revised BMP Manual for the State of North Carolina (NC DENR).

National Asphalt Pavement Association—www.hotmix.org

Interlocking Concrete Pavement Institute—www.icpi.org

Portland Cement Association—www.cement.org

National Ready Mixed Concrete Association—www.nrmca.org

BMP Database—www.bmpdatabase.org

Low Impact Development Center—www.lowimpactdevelopment.org

Center for Watershed Protection—www.cwp.org; www.stormwatercenter.net

Stormwater Manager's Resource Center—www.sustainable.org/environment/water/319-stormwater-managersresource-center-smrc

EPA Sites

Managing Wet Weather with Green Infrastructure—Permeable Pavement—EPA—water.epa.gov/infrastructure/greeninfrastructure/index.cfm

National Pollutant Discharge Elimination System Stormwater Menu of BMPs—PA—water.epa.gov/polwaste/npdes/swbmp/

Experimental Stormwater Parking Lot at Edison Laboratory—EPA—http://www.epa.gov/greeningepa/stormwater/edison_parking_lot.htm

Video: http://www.epa.gov/research/video/parking-lot/

University Research

Stormwater Center at University of New Hampshire—www.unh.edu/unhsc/

North Carolina State Stormwater Engineering Group—bae.ncsu.edu/stormwater/

Villanova University Stormwater Partnership—http://www3.villanova.edu/vusp/

PICP Resources

Interlocking Concrete Pavement Institute—www.icpi.org

Interlocking Concrete Pavement Institute (ICPI) (2009)."PICP Installer Technical Certificate Course Student Manual," 1st Ed., Interlocking Concrete Pavement Institute (ICPI), Herndon, VA, 1–148 pages. A manual for contractors attending ICPI permeable interlocking concrete pavement training courses.

(2008). "Performance Evaluation of Permeable Pavement and a Bioretention Swale." Seneca College, King City, Ontario (2005 to 2008 reports), Toronto & Region Conservation Authority, Downsview, Ontario, Canada. Comprehensive reports on three years of water quantity and pollutant monitoring a heavily used PICP parking lot near Toronto, Ontario.

"Permeable Design Pro" software (2009), Interlocking Concrete Pavement Institute, Herndon, VA. Enables users to conduct structural and hydrological design of permeable pavements.

Smith, David R. (2006). "Permeable Interlocking Concrete Pavements—Selection, Design, Construction, Maintenance," 3rd Ed., Interlocking Concrete Pavement Institute (ICPI), Herndon, VA, 1–48. A comprehensive treatment of permeable interlocking concrete pavement for design professionals.

Porous Asphalt Resources

National Asphalt Pavement Association—www.hotmix.org

"Porous Asphalt Pavements—Information Series 131," National Asphalt Pavement Association (2003) includes information on porous asphalt such as open-graded friction course information.

"University of New Hampshire Design Specifications for Porous Asphalt Pavement and Infiltration Beds," University of New Hampshire Stormwater Center, July 2007.

National Asphalt Paving Association's 2008 edition of Information Series 131 "Porous Asphalt Pavements for Stormwater Management, Design, Construction and Maintenance Guide—for specification reference for porous asphalt.

Pervious Concrete Resources

Pervious Concrete—www.PerviousPavement.org—One stop shopping for all things about pervious concrete with featured webinars.

Portland Cement Association—www.cement.org

National Ready Mixed Concrete Association—www.nrmca.org

Pacific Southwest Concrete Alliance—www.concreteresources.net

RMC Research and Education Foundation—www.rmc-foundation.org—featuring links to several research reports pertaining to pervious concrete (maintenance, durability, infiltration capabilities, etc.). Also has link to Dr. Heather Brown's Pervious Concrete Research Compilation, June 2008, a bibliography of all research related to pervious concrete.

Tennis, P. D., Leming, M. L., and Aker, D. H. (2008). "Pervious concrete Pavements." 5th Ed., Portland Cement Association and National Ready Mix Concrete Association.

American Concrete Institute (ACI) Committee 522R-06: Report on Pervious Concrete and 522.1-08: Specification for Pervious Concrete.

American Concrete Institute's ACI 522.1-08 "Specification for Pervious Concrete Pavement"—for specification reference for pervious concrete.

National Ready Mixed Concrete Association's Pervious Concrete Contractor Certification, NRMCA Publication #2PPCRT, 2005—for reference related to construction of pervious concrete.

C.5   References

American Concrete Institute (ACI) (2006). *ACI 522R—10 Report on Pervious Concrete,* American Concrete Institute ACI Technical Committee 522, Document 522R-06, Farmington Hills, MI.

American Concrete Institute (ACI) (2008). *ACI522-1-08 Specification for Pervious Concrete Pavements*, American Concrete Institute ACI Technical Committee 522, Document 522R-06, Farmington Hills, MI.

American Society for Testing and Materials (ASTM) (2008). *ASTM C1688/C1688M—08 Standard Test Method for Density and Void Content of Freshly Mixed Pervious Concrete*, ASTM International, West Conshohocken, PA.

American Society for Testing and Materials (ASTM) (2009a). *ASTM C1688/C1688M—08,* ASTM International, West Conshohocken, PA, <http://www.astm.org/Standards/C1688.htm> (Dec. 12, 2009).

American Society for Testing and Materials (ASTM) (2009b). *ASTM WK23367—New Test Method for Evaluating the Surface Durability Potential of a Pervious Concrete Mixture*, ASTM International, West Conshohocken, PA, <http://www.astm.org/SNEWS/JA_2009/c0949_ja09.html> (Dec. 12, 2009).

American Society for Testing and Materials (ASTM) (2009c). *ASTM C1701 C1701M—09 Standard Test Method for Infiltration Rate of In Place Pervious Concrete*, ASTM International, West Conshohocken, PA.

American Society for Testing and Materials (ASTM) (2013). *ASTM C1781/C1781M—13 Standard Test Method for Surface Infiltration Rate of Permeable Unit Pavement Systems*, ASTM International, West Conshohocken, PA.

Crouch, L. K., Cates, M. A., Dotson, V. J., Honeycutt, K. R., and Badoe, D. A. (2003). "Measuring the Effective Air Void Content of Portland Cement Pervious Pavements." *Cement, Concrete and Aggregates*, 25(1), 16–20.

Haselbach, L. (2010). "Pervious Concrete Testing Methods." Proc., *American Society of Civil Engineers (ASCE) Low Impact Development Conference,* San Francisco, CA.

Montes, F., Valavala, S., and Haselbach, L. M. (2005). "A New Test Method for Porosity Measurements of Portland Cement Pervious Concrete." *Journal of ASTM International*, 2(1) .

Montes, F., and Haselbach, L. M. (2006). "Measuring Hydraulic Conductivity in Pervious Concrete." *Environmental Engineering Science*, 23(6), 956–965.

D Acronyms and Glossary

D.1 Acronyms

AASHTO	American Association of State Highway and Transportation Officials
ACI	American Concrete Institute
ASCE	American Society of Civil Engineers
ASTM	ASTM International, formerly American Society for Testing and Materials
ATPB	Asphalt treated permeable base
BMP	Best management practice
CBR	California Bearing Ratio
CSO	Combined sewer system overflow
CPTech Ctr	National Concrete Pavement Technology Center
CSA	Canadian Standards Association
CN	Curve number
DGHMA	Dense graded hot mix asphalt (opposite of porous asphalt). Typically referred to as asphalt or hot mix asphalt (HMA)
DOT	Department of Transportation
EPA	US Environmental Protection Agency
EAL	Equivalent axle load
ESAL	Equivalent single axle load
EWRI	Environmental and Water Resources Institute of ASCE
GIS	Geographic information systems

HMA	Hot mix asphalt
HSG	Hydrological soil group
ICPI	Interlocking Concrete Pavement Institute
IgCC	International Green Construction Code
LCA	Life cycle analysis or assessment
LCCA	Life cycle cost analysis
LID	Low impact development
MS4s	Municipal Separate Storm Sewer System
NAPA	National Asphalt Paving Association
NCDENR	North Carolina Department of Environment and Natural Resources
NPDES	The National Pollutant Discharge Elimination System (Federal law with the program overseen by the EPA.)
NRMCA	National Ready Mix Concrete Association
NYSDEC	New York State Department of Environmental Conservation
OGFC	Open-graded friction course
PCA	Portland Cement Association
PFC	Permeable friction course
PICP	Permeable interlocking concrete pavement
SCM	Stormwater control measures
SN	Structural number
SWMM	Stormwater management model
TKN	Total Kjeldahl nitrogen
TMDL	Total maximum daily load
TRB	Transportation Research Board
TSS	Total suspended solids

UNHSC	University of New Hampshire Stormwater Center
UIC	Underground Injection Control
UWRRC	Urban Water Resources Research Council of the Environmental and Water Resources Institute of ASCE
WMA	Warm mix asphalt

D.2 Glossary

Aggregate: Crushed stone or gravel used for jointing, bedding, base and subbase materials and in concrete mixes

Angularity: The sharpness of edges and corners of sand and aggregate particles

Aquifer: A porous water-bearing geologic formation that yields water for consumption

Aspect ratio: The longest overall length of a paver divided by its thickness (Example: a 100 mm [4 in.] wide by 200 mm [8 in.] long by 80 mm [3.15 in.] thick paver has an aspect ratio of 2.5.); compare to plan ratio

Asphalt pavement: For the purposes of this document, pavement made with asphalt as the binder (See also concrete pavement.)

Asphalt treated permeable base (ATPB): Base material with 3% to 4% liquid asphalt added for stability

Base or **base course**: A material of a designed thickness placed under the surface wearing course of paving and bedding courses. It is placed over a subbase or a subgrade to support the surface course and bedding materials and for permeable pavements may also serve as a reservoir bed or course. A base course can be compacted aggregate, or asphalt stabilized aggregate, asphalt, or concrete. It is considered a base when there is only one layer between the pavement and subgrade.

Bedding course: A layer of coarse, crushed and washed stone screeded smooth as bedding for the pavers. This material generally conforms to the grading requirements of ASTM No. 8 stone. This screeded permeable aggregate layer is generally 50 mm (2 in.) thick.

Best Management Practice (BMP): A structural or non-structural device/practice designed to infiltrate, temporarily store or treat stormwater runoff in order to reduce pollution and flooding

Bioretention: A stormwater management practice that uses soils and vegetation to treat pollutants in urban runoff and to encourage infiltration of stormwater into the ground

Bioretention basins: Landscaped depressions or shallow basins used to slow and treat on-site stormwater runoff. Stormwater is directed to the basin where it is treated by physical, chemical, and biological processes. The slowed, cleaned water infiltrates native soils or is directed to nearby stormwater drains or receiving waters. PICP overflow can drain into such basins.

California Bearing Ratio (CBR): A test method and result that renders an approximation (expressed as a percent) of the bearing strength of soil compared to that of a high quality, compacted aggregate base. The test defines the ratio of: (1) the force per unit area required to penetrate a soil mass with a 19 cm^2 (3 in.^{2}) circular piston (approximately 51 mm [2 in.]) diameter] at the rate of 1.3 mm/min (0.05 in./min), to (2) that required for corresponding penetration of a standard material. The ratio is usually determined at 2.5 mm (0.1 in.) penetration, although other penetrations are sometimes used. See ASTM D1883 or AASHTO T193.

Capillary barrier: A layer or course of coarser open-graded aggregate to prevent moisture from entering an overlying layer through capillary action

Cation: A positively charged atom or group of atoms in soil particles that, through exchange with ions of metals in stormwater runoff, enable those metals to attach themselves to soil particles

Chamfer: A 45° beveled edge around the top of a paver unit usually 2 to 6 mm (0.08 to 0.24 in.) wide; it allows water to drain from the surface, facilitates snow removal, helps prevent edge chipping and delineates the individual paving units

Choke course: A layer of aggregate placed or compacted into the surface of another layer to provide stability and a smoother surface; the particle sizes of the choke course are generally smaller than those of the surface into which it is being pressed

Clay soils: 1. (Agronomy) Soils with particles less than 0.002 mm (8 in.) size
          2. A soil textural class
          3. (Engineering) A fine-grained soil with more than 50% passing the No. 200 sieve with a high plasticity index in relation to its liquid limit, according the Unified Soil Classification System

Coarse aggregate: Typically that portion of an aggregate retained on the 4.75 mm (No. 4) sieve (See also **fine aggregate**.)

Combined sewer system overflow (CSO): Conveyance of storm and sanitary sewage in the same pipes

Compaction: The process of inducing close packing of solid particles such as soil, sand, aggregate or a combination thereof

Compressive strength: The measured maximum loading resistance expressed as force per unit cross-sectional area such as pounds per square inch (PSI) or newtons per square millimeter (megapascals)

Concrete pavement: Technically concrete is a composite material with aggregate and a binder, but for the purposes of this document, concrete pavement is pavement made with cement as the binder (See also **asphalt pavement**.)

Concrete pavers: Precast concrete units meeting the requirements of ASTM C936 or CSA A231.2, but for the purpose of this standard, having an aspect ratio of 3 or less

Course: A row of pavers or a layer in a permeable pavement system

Crushed stone: A product used for pavement bases/subbases made from mechanical crushing of rocks, boulders or large cobblestones at a quarry; all faces of each aggregate have well-defined edges resulting from the crushing operation

Curve number (CN): A numerical representation of a given area's hydrological soil group, plant cover, impervious cover, interception and surface storage. The US Soil Conservation Service (SCS) now the Natural Resource Conservations Service (NRCS) originally developed the concept. A curve number is used to estimate runoff volume from rainfall depth.

Dense-graded aggregate base: A compacted crushed stone base whose gradation yields very small voids between the particles with no visible spaces between them. Most dense-graded bases have particles ranging in size from 38 mm (1.5 in.) or 19 mm (0.75 in.) down to fines passing the 0.075 µm sieve (No. 200) sieve.

Density: The mass per unit volume

Detention pond or structure: The temporary storage of stormwater runoff in an area

Edge restraint: A curb, edging, building, or other stationary object that contains the bedding course and pavers or pavement

Equivalent single axle loads (ESALs): Summation of equivalent 80 kN (18,000 lbf) single axle loads used to combine mixed traffic to a design traffic load for the design period; also expressed as Equivalent Axle Loads or EALs

Erosion: The process of wearing away of soil by water, wind, ice, and gravity; detachment and movement of soil particles

Evapotranspiration: The return of moisture to the atmosphere from the evaporation of water from soil and transpiration from vegetation

Exfiltration: The downward movement of water through a permeable pavement system into the soil beneath

Fine aggregate: Typically sand (See also **coarse aggregate**.)

Fines: Silt and clay particles in a soil, generally those smaller than the No. 200 or 0.075 mm sieve

First flush: The initial portion of a rainstorm that picks up high concentrations of accumulated pollutants, which are usually due to antecedent dry weather conditions which create an accumulation of pollutants on pavements

Flexible pavement: A pavement structure which maintains intimate contact with and distributes loads to the subgrade. The base course materials rely on aggregate interlock, particle friction, and cohesion for stability. Asphalt is a flexible pavement. (See also **rigid pavement**.)

Filter course: A layer in a permeable pavement system for water quality or hydrological purposes, usually made of sand. Its location varies depending on the system design.

Frost action: Freezing and thawing of moisture in pavement materials and the resultant effects on them

Geotextiles or **geotextile fabrics**: Woven or non-woven fabrics used for separation, reinforcement, or drainage between pavement layers or neighboring soils, etc.; AASHTO M288 is a common reference for selection criteria.

Gradation: Soil, sand, or aggregate distributed by mass in specified particle-size ranges. Gradation is typically expressed in percent of mass of sample passing a range of sieve sizes. See ASTM C136.

Grade: (*noun*) The slope of finished surface of an excavated area, base, or pavement usually expressed in percent
         (*verb*) To finish the surface of same by hand or with mechanized equipment

Gravel: Rounded or semi-rounded particles of rock that will pass a 75 mm (3 in.) and be retained on a 4.75 mm (0.19 in.) (No. 4) sieve which naturally occurs in streambeds or riverbanks that have been smoothed by the action of water. A type of soil as defined by the Unified Soil Classification System having particle sizes ranging from the 4.75 mm (0.19 in.) (No. 4) sieve size and larger.

Herringbone pattern: A pattern where joints are no longer than the length of 1.5 pavers. Herringbone patterns can be 45° or 90° depending on the orientation of the joints with respect to the direction of the traffic.

Hotspot: A land use that generates highly contaminated runoff with concentrations higher than those typical to stormwater

Hydrological soil group (HSG): The soils classification system developed by the US Soil Conservation Service (now the Natural Resource Conservation Service) that categorizes soils into four groups, A through D, based on runoff potential. A soils have high permeability and low runoff whereas D soils have low permeability and high runoff.

Impervious cover: Any surface in the built environment that prohibits percolation and infiltration of rainwater into the ground; a term for most pavements and roofs

Infiltration rate: The rate at which stormwater moves into the top surface of the pavement or ground measured in inches or centimeters per hour.

Interlocking concrete pavement: A system of paving consisting of discrete, hand-sized paving units with either rectangular or dentated shapes manufactured from concrete. Either type of shape is placed in an interlocking pattern, compacted into coarse bedding sand, the joints filled with sand and compacted again to start interlock. The paving units and bedding sand are placed over an unbound or bound aggregate layer. Also called concrete block pavement.

Joint: The space between concrete paving units typically filled with small-sized open-graded aggregate. Or separation of a concrete pavement slab from the neighboring slab.

Jointing aggregate: Small sized aggregates swept into the openings between the pavers. The aggregate size varies the joint width. Typical sizes are ASTM No. 8, No. 89, or No. 9 stone.

Karst geology: Regions of the earth underlain by carbonate rock typically with sinkholes and/or limestone caverns

Layer coefficient: From the 1993 AASHTO pavement design procedure; a dimensionless number that expresses the material strength per 25 mm (1 in.) of thickness of a pavement layer (surface, base or subbase)

Laying pattern: The sequence of placing pavers where the installed units create a repetitive geometry. Laying patterns may be selected for their visual or structural benefits.

Life cycle analysis/assessment (LCA): A method of estimating environmental and resource impacts of a product, process, or system

Life cycle cost analysis (LCCA): A method of calculating all costs anticipated over the life of the pavement including construction costs. Discounted cash-flow methods are generally used, typically with calculation of present worth and annualized cost. Factors that influence the results include the initial costs, assumptions about maintenance and periodic rehabilitation, pavement user and delay costs, salvage value, inflation, discount rate, and the analysis period. A sensitivity analysis is often performed to determine which variables have the most influence on costs.

Lift: During placement of a pavement system, a lift is a layer or portion of a layer that is placed and worked on separately from another layer

Low impact development (LID): Developmental design that tries to mimic the natural hydrologic cycle more closely

Mechanical installation: The use of machines to lift and place layers of pavers on screeded sand in their final laying pattern. It is used to increase the rate of paving.

Mechanistic design: Analysis of structural response of applied loads through modeling of stresses and strains in a pavement structure.

Modified Proctor Test: A variation of the Standard Proctor Test used in compaction testing which measures the density-moisture relationship under a higher compaction effort. (See also **Proctor density**.)

Modulus of elasticity or **elastic modulus**: The ratio of stress to strain for a material under given loading conditions

Municipal Separate Storm Sewer System (MS4s): A conveyance or system of conveyances that is: owned by a state, city, town, village, or other public entity that discharges to waters of the US; designed or used to collect or convey stormwater (including storm drains, pipes, ditches, etc.); not a combined sewer; and not part of a Publicly Owned Treatment Works (sewage treatment plant).

Observation well: A perforated pipe inserted vertically into a base/subbase used to monitor its contents

One year storm: A rainfall event that has a 100% probability of it, or a larger storm, occurring in a given year

One hundred-year storm: A very infrequent rainfall event that has a 1% chance of it or a larger storm occurring in a given year

Open-graded base: Generally a crushed stone aggregate material used as a pavement base/subbase that has no fine particles in it. The void spaces between aggregate can store water and allow free drainage from the base.

Open-graded Friction Course (OGFC): A top layer of pavement that is has large enough voids for water to infiltrate placed over an impermeable pavement layer. The flow then moves horizontally to the sides of the pavement. It can have noise reduction, hydrological, and safety benefits. (See also **permeable friction course [PFC]**.)

Outlet: The point at which water is discharged from an open-graded base/subbase through pipes into a stream, lake, river or storm sewer

Pavement structure: A combination of subbase, base course, and surface course placed on a subgrade to support traffic loads and distribute it to the roadbed

Peak discharge rate: The maximum short term flow from a detention or retention pond, open-graded base, pavement surface, storm sewer, stream, or river usually related to a specific storm size event

Performance period: The period of time that an initial pavement structure will last before requiring rehabilitation. The performance period is equivalent to the time elapsed as a new, reconstructed, or rehabilitated pavement structure deteriorates from its initial serviceability to its terminal serviceability

Perforated underdrain: A perforated piping system to carry flow from the reservoir layer of a permeable pavement system. Its vertical location varies depending on design and conditions such as retention or detention, water quality issues and frost depths (See also **underdrain**.)

Permeability: The rate of water movement through a soil column. Most commonly under saturated conditions (saturated hydraulic conductivity)

Permeable friction course (PFC): A top layer of pavement that is has large enough voids for water to infiltrate placed over an impermeable pavement layer. The flow then moves horizontally to the sides of the pavement. It can have noise reduction, hydrological and safety benefits. (See also **open-graded friction course (OGFC)**.)

Permeable interlocking concrete pavement (PICP): A paving system consisting of discrete, hand-sized paving units with rectangular or dentated shapes manufactured from concrete. Either type of shape is placed in an interlocking pattern, compacted into a highly permeable bedding layer consisting of small aggregate, the joints filled with a highly permeable aggregate compacted again to start interlock. The paving units and bedding are placed over a highly permeable open-graded base typically 200 mm (4 in.) thick of aggregates ranging in size from 20 mm down to 3 mm (0.79 in. down to 0.12 in.). This layer is placed over an open-graded subbase consisting of larger sized aggregates, an unbound or occasionally a bound layer. Sand is not used within the pavement structure.

Permeable pavement: A surface with penetrations capable of passing water and supporting pedestrians and vehicles. Typical examples are pervious concrete, porous asphalt, permeable interlocking concrete pavement, and permeable or grid pavers. (See also **porous pavement**.)

Permeable pavement system: A permeable pavement and the underlying layers/courses and features for support, storage, etc.

Pervious concrete: A type of permeable pavement made of cement concrete

Pervious or **permeable surfaces/cover**: Surfaces that allow the infiltration of rainfall such as vegetated areas

Porosity: Volume of voids divided by the total volume

Porous asphalt: A type of permeable pavement made of asphalt concrete

Porous pavement: A surface full of pores capable of supporting pedestrians and vehicles. (See also **permeable pavement**.)

Pretreatment: BMPs that provide initial storage and filtering pollutants before they enter another BMP

Proctor density: The Proctor Compaction Test is a laboratory method of experimentally determining the optimal moisture content at which a given soil type will become most dense and achieve its maximum dry density.

Reservoir bed/layer/course: The layer in a permeable pavement system for detention or retention of water (See also **storage bed**.)

Retention pond or **structure**: Collects runoff and allows for infiltration into the soils below or long term storage instead of detained discharge (See also **detention pond or structure**.)

Rigid pavement: Concrete is a rigid pavement (See also **flexible pavement**.)

Runoff coefficient: Ratio of surface runoff to rainfall expressed a number from 0 to 1

Sailor course: A paver course where lengths of rectangular pavers are laid parallel (lengthwise) to the edge restraint

Sand: 1. (Agronomy) A soil particle between 0.05 and 2.0 mm (0.01 and 0.08 in.) in size
    2. A soil textural class
    3. (Engineering) A soil larger than the No. 200 (0.075 mm [0.01 in.]) sieve and passing the No. 4
      (4.75 mm [0.19 in.]) sieve, according to the Unified Soil Classification System (USCS)

Sediment: Soils transported and deposited by water, wind, ice, or gravity

Sheet flow: The laminar movement of runoff across the surface of the landscape

Silt: 1. (Agronomy) A soil consisting of particles sizes between 0.05 and 0.002 mm (0.01 and 0.08 in.)
    2. A soil textural class
    3. (Engineering) A soil with no more than 50% passing the No. 200 (0.075 mm [0.01 in.]) sieve that has
      a low plasticity index in relation to the liquid limit, according to the Unified Soil Classification System

Serviceability: The ability of the pavement to serve the type of traffic (trucks) which use the facility. The primary measure of serviceability is the Present Serviceability Index (PSI), ranges from 0 (very poor road) to 5 (perfect road).

Storage bed/layer/course: The layer in a permeable pavement system for detention or retention of water (See also **reservoir bed**.)

Structural Number (SN): A calculation used by AASHTO to assess the structural capacity of a pavement to handle loads based on ESALs and soil subgrade strength

Swale: A linear topographic depression that conveys runoff

Subbase: The layer or layers of specified or selected material of designed thickness placed on a subgrade to support a base course. Aggregate subbases are typically made of stone pieces larger than that in bases. (When there is only one layer between the pavement and subgrade, it can be called the base.)

Subgrade: The soil upon which the pavement structure and shoulders are constructed

Time of concentration: The time runoff takes to flow to a drainage area's most distant point to the point of drainage

Total maximum daily load (TMDL): A term in the US Clean Water Act describing the maximum amount of a pollutant a body of water can receive within a certain period of time without significantly impairing the water quality or health of the existing aquatic ecosystem

Treated base: An aggregate base with cement, asphalt or other material added to increase its structural capacity

Underdrain: A piping system to carry flow from the reservoir layer of a permeable pavement system. Its vertical location varies depending on design and conditions such as retention or detention, water quality issues and frost depths. (See also **perforated underdrain**.)

Void ratio: Volume of voids around the aggregate divided by the volume of solids

D.3 References

American Association of State Highway and Transportation Officials, *AASHTO Guide for the Design of Pavement Structures*, Washington D. C., 1993.

American Association of State Highway and Transportation Officials (AASHTO), *Standard Specification for Geotextile Highway Applications (M-288-06)*, Washington DC, 2006.

American Association of State Highway and Transportation Officials (AASHTO), *Standard Method for Testing Resilient Modulus (T307)*, Washington DC, 2006.

American Association of State Highway and Transportation Officials (AASHTO), *Standard Method of Test for Moisture-Density Relations of Soils Using a 4.54 kg (10 lb) Rammer and a 457 mm (18 in.) Drop (T180)*, Washington DC, 2009.

American Association of State Highway and Transportation Officials (AASHTO), *Standard Method of Test for the California bearing Ratio (T-193-99)*, Washington DC, 1999.

ASTM International, *Standard Specification for Concrete Aggregates (C33/33M-08), Annual Book of ASTM Standards*, Conshohocken, Pennsylvania, 2008.

ASTM International, *Standard Specification of Soils for Engineering Purposes (Unified Soil Classification System (D2487-00), Annual Book of ASTM Standards*, Conshohocken, Pennsylvania, 2000.

ASTM International, *Standard Test Method for Resistance to Degradation of Small-Size Coarse Aggregate by Abrasion and Impact in the Los Angeles Machince (C131), Annual Book of ASTM Standards*, Conshohocken, Pennsylvania, 2006.

ASTM International, *Standard Specification for Solid Concrete Interlocking Paving Units (C939), Annual Book of ASTM Standards*, Conshohocken, Pennsylvania, 2012.

ASTM International, *Standard Test Method for California Bearing Ratio (CBR) of Laboratory Compacted Soils (D1883), Annual Book of ASTM Standard*s, Conshohocken, Pennsylvania, 2007.

ASTM International, *Standard Practice for Classification of Soils for Engineering Purposes (Unified Soil Classification System) (D2487), Annual Book of ASTM Standards*, Conshohocken, Pennsylvania, 2006.

ASTM International, *Standard Test Method for Resistance R-Value and Expansion Pressure of Compacted Soils (D2844), Annual Book of ASTM Standards*, Conshohocken, Pennsylvania, 2007.

Canadian Green Building Council, LEED *Green Building Rating System for New Construction and Major Renovations* (LEED-Canada NC version 1.0), Ottawa, Ontario, December 2004.

Canadian Standards Association, *Precast Concrete Pavers (A231.2)*, CSA Standards, Rexdale, Ontario, 2006.

Environment Canada, *Canadian Climate Normals*, Atmospheric Environment Service, Environment Canada, Toronto, Ontario, Canada, 2010.

Giroud, J. P., "Review of Geotextile Filter Criteria" in *Proceedings* of the 1st Indian Geotextiles Conference on Reinforced Soil and Geotextiles, Bombay, India, pp. 1–6., 1988.

Giroud, J. P., Granular Filters and Geotextile Filtyers. *Proceedings* of Geofilters 1996, Montreal, Canada, pp. 565–680., 1996.

Hershfield, D. M., *Rainfall Frequency Atlas of the Unites States* (for durations from 30 minutes to 24 hours and return periods from 1 to 100 years), <http://www.nws.noaa.gov/oh/hdsc/PF_documents/TechnicalPaper_No40.pdf>, US Department of Commerce, Weather Bureau, Technical Paper No. 40, Washington DC, 1961.

Interlocking Concrete Pavement Institute, *ICPI Teh Spec 1–Glossary of Terms Used in the Production, Design, Construction, and Testing of Interlocking Concrete Pavement*, Herndon, Virginia, 2006.

Miller, J. F., Frederick, R. H. and Tracey R. J., *Precipitation-frequency Atlas of the Western United States*, Volumes I–XI, Atlas 2, <http://www.nws.noaa.gov/oh/hdsc/noaaatlas2.htm>, US Department of Commerce, National Weather Service, National Ocean and Atmospheric Administration, Silver Spring, Maryland, 1973.

NRCS, *National Engineering Handbook*, Part 630 Hydrology, Natural Resource Conservation Service, US Department of Agriculture, Washington D. C., May 2007.

Smith, D. R., *Permeable Interlocking Concrete Pavements*, 4th Edition, Interlocking Concrete Pavement Institute, Herndon, VA, 2011.

U.S. Green Building Council, *LEED Green Building Rating System for New Construction and Major Renovations* (LEED-NC) Version 3, Washington, D. C. April 27, 2009.